British Coal

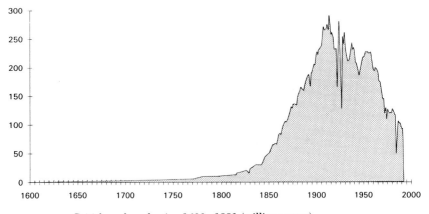

British coal production 1600–1991 (million tonnes)

British Coal

Prospecting for privatization

Charles Kernot

WOODHEAD PUBLISHING LIMITED

Cambridge England

Published by Woodhead Publishing Limited, Abington Hall,
Abington, Cambridge, CB1 6AH, England

First published 1993, Woodhead Publishing Limited

© *C P H F Kernot, 1993*

British Library Cataloguing in Publication Data
A catalogue record for this book is available from the British Library

ISBN 1 85573 120 7

Designed by Andrew Jones (text) and Chris Feely (jacket)
Typeset by Best-set Typesetter Ltd, Hong Kong
Printed by Foister & Jagg Ltd, Cambridge, England

For Oma and Opa

Contents

CONTENTS

Preface

*I do not see the government's task as being to try to plan the
future shape of energy production and consumption. It is not
even primarily to try to balance UK demand and supply for
energy. Our task is rather to set a framework which will
ensure that the market operates in the energy sector with
a minimum of distortion and that energy is produced and
consumed efficiently.*

Speech by the Minister for Energy, Nigel Lawson MP, June
1982.

On 13 October 1992 British Coal announced its intention to close 31 of its
50 operating collieries with the loss of some 30 000 jobs. Whilst the
decision to close three of the collieries had already been detailed, and despite
the large number of warnings (from both sides of the political divide) that a
major decision was imminent, it still shocked the nation. Even now, there
remains a feeling of disbelief that the industry on which Britain's prosperity
was founded should be reduced to such a pitiable state.

This book charts the rise of the British coal industry from its earliest
beginnings and follows it through the industrial revolution to the peak of
output in 1913. Then, to the end of Part I, it covers the long decline to the

situation that faced the mines on 13 October 1992. In order to put this into an international context, Part II covers the worldwide market in coal and shows Britain's place in that market. Part III, finally, looks at the current energy market in the UK and provides some guide-lines as to how it might change over the coming years. This is, necessarily, a very difficult subject to cover and it has been possible to look at it only in the broadest terms. With privatization still intended (albeit 10 years later than originally planned), thoughts and ideas about how this could be achieved and, more importantly, how much money the government might raise are included before a final chapter draws all of the threads together and updates the reader for the major events which took place between the closure announcement and the publication of the White Paper.

Many readers may not like the conclusions drawn throughout this book – whether their political leanings are to the left or the right. Indeed, throughout its existence it has become more and more apparent to me that the coal industry has suffered from both bad management and an intransigent work-force. The forty-six years since nationalization have seen no let up in this picture of incompetence and it is to the discredit of the Conservative Party under Margaret Thatcher and the idealized rhetoric of Arthur Scargill that the sorry situation of the October closures came to pass.

However, there is no point in thinking about what might have been – the current state of the energy market in the UK is what Britain's coal mines have to live with and because of their high costs there is little alternative but to conclude that many of the collieries had to close. Alternatives do exist, but they would have involved a complete rethink of Conservative free market ideology and dogma and would have set back the picture of Britain as a stable political environment in which to do business by many decades. The other alternative would have required the government to summon up the courage to pursue the opening up of the energy market to its privatized potential. This too, it has so far failed to do.

In British Gas, privatization created a monolith that is fighting hard to keep its monopoly. In the electricity generation industry in England and Wales an oligopoly was created – one stronger than the other but both capable of exerting pressure on the market price of their product. And despite the presence of Nuclear Electric there should be no surprise that the electricity suppliers (Regional Electricity Companies or RECs) chose to diversify their source of supply away from this oligopoly, thereby stimulating the dash for gas. Is it any wonder, therefore, that the generators should look to diversify their own production and reduce costs? Indeed, it is not; and they have achieved this both through the construction of their own gas-fired plant and, more importantly for British Coal, through the purchase of much cheaper coal from overseas.

Unfortunately for British Coal it was not offered a range of contracts with the generators in advance of their privatization. This was a major error on the government's part as the termination of all contracts on a single day lies at the heart of British Coal's problems. None but the most foolhardy business executive would risk all of a company on a single contract with no guarantee of renewal; but such was the contract forced on British Coal. Whilst it must be admitted that the end of the contract would free the energy market from a major distortion there are still many areas where the market in energy is not entirely free. And, although these have made it possible to maintain the status quo, the termination of these distortions in fits and starts associated with massive programmes of investment or closure has added to the instability of the market.

In many cases the opening of one new area of supply had led to a decline in another. In British Coal's case (and in the government's defence it may not have been possible to foresee) the company has had to face two new areas of competition – cheap foreign imports and gas – within the period it takes to dig a new mine. Such were the flaws in the government's privatization strategy – the market was indeed driving energy policy but the speed attained on an unknown road did not allow for any reversal in the case of a wrong turn.

I would like to take this opportunity to thank all of those people who have helped and supported me through this endeavour. First must be my girl-friend, Emma and her parents, Evelyn and Alan, who have all provided me with much support. Richard Orme (no relation to Stan Orme MP, I believe) and Robert Thomson also deserve a mention for helping to keep me sane, whilst all members of the Trade and Industry Select Committee, who showed their concern for the plight of the industry, proved to me that there is compassion in politics. I would also like to take this opportunity to acknowledge the sources of the Charts and Tables that I have used. Where there were only one or two sources these have been acknowledged within the text. Otherwise readers are referred to the bibliography at the end of the volume. Finally, I would also like to thank all of those people who supplied me with information, and I apologize if they do not find it in the text as the constraints of space have forced me to keep within the specific remit of the title.

Charles Kernot

Part

I

The history of British coal mining

Chapter

1

Up to the industrial revolution . . .

A QUESTION OF DATES

If mechanization is used as a strict definition of industrialization then the industrial revolution of the British coal mining industry did not start in earnest until the beginning of the twentieth century. The first real industrial advance was the introduction of mechanical cutting tools in the late nineteenth century but by 1900 they only accounted for 1% of the total of 228.4 million tonnes of coal mined in the country. The further developments of power loading and mechanical conveying did not take hold until the 1920s and 1940s respectively; a diabolical state of affairs, which left the British coal mining industry at the bottom of the world productivity table.

The delay in the implementation and introduction of new techniques and equipment was one of the reasons behind Britain's poor performance in the years between the two World Wars. Whilst the productivity of a mine, of necessity, declines with age, a move to mechanical cutting and power loading can increase the output per manshift considerably. However, one of the causes of Britain's poor performance was that so many of the mines had already been planned and dug before these technological advances were made. As a result the cost of altering the existing mine was too great to overcome, whilst the

3

availability of relatively cheap labour encouraged the mine owners to conclude that the extra expense could not be justified.

Looking further back, to the industrial revolution proper, the development of the pumping engine around the turn of the eighteenth century meant that mines could be sunk below the water table with greater efficiency and with less fear of flooding. However, this was of little consequence at the time as mines were still relatively shallow and the pumps inefficient. In any case the demand for coal had not started to increase to a level which required larger, deeper mines and which, in turn, would lead to the use of a large number of pumping engines. One of the difficulties in stimulating demand was the inadequate transport network which forced up the price of coal away from the pit-head. It was not until the 1750s and 1760s that canals, and other advances in transportation, helped to reduce the price of coal to levels which started to stimulate demand. These years also saw an increase in coal exports from England and Wales to levels not seen before and began the start of a major export industry for the country (see Figure 1.1).

As far as mining techniques were concerned the introduction of improved ventilation through the use of brattices to guide fresh air through the mines started in the 1750s. The decade also saw the introduction of pit ponies for the first time and it is strange to consider that they were still employed in a handful of collieries in the 1970s. Whilst these improvements cannot be considered as revolutionary they certainly helped mines to increase output to match the increase in demand which was occasioned by the industrial revolution in other sectors of the economy. Overall, therefore, this chapter covers the start of coal mining in Britain from the very earliest times up until the middle of the eighteenth century, despite the fact that the century as a

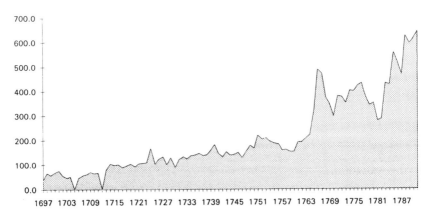

1.1 Coal exports from England and Wales 1697–1791 (000t)

whole saw a fivefold increase in coal production. It increased by a much more significant 22.5 times in the nineteenth century.

THE EARLY DAYS

Demand

Coal has been used in Britain at least since Roman times and probably from long before. The earliest details come from Hadrian's Wall where Roman troops are thought to have burned coal in order to keep out the Scottish cold over the long winter nights, and from Bath where coal provided the sacred flame in the Temple of Minerva. Nevertheless, mining is unlikely to have been carried out on an organized basis until the eleventh or twelfth centuries and the initial gathering of coal probably took place where seams were exposed at the surface. The first concerted efforts to extract coal were along the north-east coast where the coal measures were exposed to the erosive effects of the North Sea. This brought them to the attention of local inhabitants who burned cheap coal in preference to the more expensive wood. However, the difficulty and expense of transport meant that there was little movement of coal from its immediate source until the thirteenth century.

Indeed, it is in the thirteenth century that coal gets its first specific mentions. Surprisingly, perhaps, there is no specific reference in the Domesday Book and nor did any of the other chroniclers of the earlier centuries give coal any mention. The first explicit reference to coal seems to have been by the Bishop of Durham in 1200 when he attempted to promote the development of coal in the Tyne valley. In 1257 Queen Eleanor was forced to flee London because of the 'fumes of the sea coals' and both Seacoal and Old Seacoal Lanes, which are to be found by Ludgate Circus in the City of London, date from the thirteenth century with the first reference in 1226.

The term 'sea coal' probably relates to the method of transport rather than the area of extraction. From the north-east coast by far the easiest method of transport to London in these times was by coaster. The roads were in a dire state and it was not until the middle of the eighteenth century that the canal building programme really got underway. Before this the price of coal in some areas tended to fluctuate with the weather as roads often became impassable when wet – the greater the amount of rain the higher the price of coal. The average cost of transport was at least four times as great by road as by canal, and could be up to 20 times more for transporting coal. Indeed, a single horse could pull as much coal in a barge as could 80 horse and carts or 400 pack horses. For these reasons the majority of coal produced until the start of the sixteenth century was used within a two mile radius of a mine, and was largely restricted to the poor who could not afford to buy wood. The preference for wood was largely related to the lack of chimneys in most houses, which often just had a single central fire. As coal fires are sooty and smelly the cleaner, and in some cases more aromatic, wood was burned despite commanding a higher price.

The main reason for London's demand for coal was simply that the forests around it had been denuded and there was little other fuel available locally at an acceptable price. One of the main reasons for the loss of the forests and consequent timber shortage was that land for pasture was worth three times more than forestry during the early 1600s. This was simply because of the greater return which could be generated from animal husbandry than from collecting wood and burning charcoal. As the demand for grain grew with the population so arable land also increased in value and by the 1660s arable land, which was being intensively cultivated, was worth three times more than pasture. Had there been insufficient coal available in London then the price differential of land is unlikely to have been so large, although the expanding population of the capital meant that agricultural produce was also important. The population of London increased fourfold between 1500 and 1600 and led to a great increase in demand for country produce. As coal could be transported by sea it could be moved right into the centre of the city and, therefore, directly to its market with little need for overland movement. As a result the city's lightermen and bargemen had something of a monopoly on the supply of coal to London for many years.

Production

It is difficult to estimate the total amount of production during the early years, and then even the period from 1765 until the middle of the nineteenth century is something of an educated guess as the first countrywide statistics were not collated until 1854. Most of the work from these early years was carried out by Professor Nef who collated as many statistics as possible for different mines across the country. He then amalgamated these with estimates of the total number of mines in operation in order to derive an accurate estimate of the total level of production. His, and other figures show that the annual production of coal probably averaged some 210 000 t in the 1550s and slowly grew to around 2.25 mt in 1660, 2.5 mt in 1700 and around 6 mt in 1770. The sixteenth century saw the start of the upturn in coal output together with the first real increase in transportation to areas other than London. The increase in demand and production is sometimes referred to as the sixteenth century coal rush (Figure 1.2).

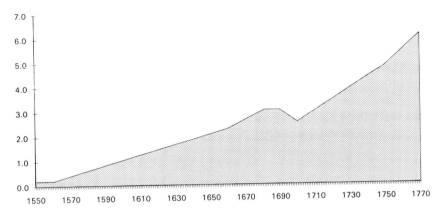

1.2 Coal production 1550–1770 (million tonnes)

The sixteenth century is also important as it shows the beginnings of the division of labour as the first full time, specialist miners are recorded. Before this the mines were small scale operations employing no more than 12 miners who were from the locality. Even so these workers will not have been employed for the full year as the need for agricultural labour during harvests meant that they were often relieved of their duties to help elsewhere on their master's estate. Indeed, the ownership of the early mines was restricted to the few landowners in the vicinity of the coast and, for many years, one of the largest of these was the Church. Only the dissolution of the monasteries in the 1530s brought these properties into wider ownership following their distribution by Henry VIII as recompense for the loyalty he had been offered by members of his court. This additionally occurred at the time that the timber famine was starting to become serious and was stimulating the search for alternative sources of energy.

It should further be considered that most of the other known areas of coal at the time were still in politically unstable countries and principalities and that there was therefore little fear of import competition. Both of the coal fields in what are now Belgium and Germany had large reserves, but the unstable politics of these regions meant that the large amounts of capital required to serve the British market could not be justified. This was despite the ease of access down the Ruhr and across the Channel. Therefore, mainland Britain was left in relative peace and security and this helped to promote the investment of capital in increasingly large scale operations. By the start of the eighteenth century coal was being mined in all areas of Britain, with the exception of the deep Kent coalfield which did not start to be exploited until the early twentieth century.

7

The mines from which the coal was extracted were nothing like the great enterprises which exist today. As mentioned above they employed probably no more than a dozen people and were either simple drift mines cut into a hill or cliff, or they were bell pits dug out from a single shaft sunk into the ground. These pits could only extract coal from a restricted area at the bottom of the shaft because of the danger of rockfalls and the lack of ventilation provided by the single entrance. Indeed, it was not until the Hartley Colliery disaster in 1862, when 204 men suffocated at the bottom of a shaft because they had no other means of escape, that it became mandatory for two shafts to be sunk in every coal mine.

Costs and prices

The use of coal in London is first documented on a consistent basis by Westminster School which kept records of the prices it paid for coal between 1585 and 1830. These prices are detailed in Figure 1.3 and show that there were often marked fluctuations in the price the school had to pay for its coal. These prices were often linked to periods of rising or falling prices generally, and the rise in price during the high inflationary period of the Napoleonic Wars is clearly shown. This price rise is related both to the rate of inflation and lack of manpower with troops fighting on the continent, but is also related to the danger of attack by French privateers who sought to prevent coal reaching London.

The first descriptions of coal mines exist from the fourteenth century when mining seems to have become an entrepreneurial operation with the coal

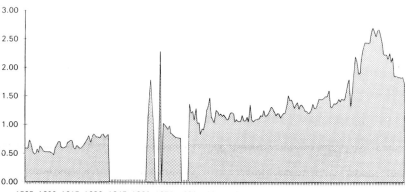

1.3 Price of coal at Westminster School 1585–1830 (£/tonne)

8

often worked for sale rather than for personal consumption. In 1316 there is a description of a mine at Cossall in Nottinghamshire being worked by nine men. In order to obtain the right to work the mine they had to pay 12d a week for each 'pickaxe' used, although they were able to obtain a reduction if they were unable to work due to flooding or gas. As production was in the region of 3 t/week the royalty payment worked out at around 4d/t or about one-third of the pithead selling price required to cover all of their costs. There are other examples of the industry from the period which include lease payments for the land from which the coal was produced. Whilst an individual may have wanted to dig his own coal to save on the cost of purchase it is unlikely that coal would just have been mined for this purpose and some must have been for sale to third parties.

Leases were a significant part of the total cost of the coal and by the seventeenth century they had been transferred into royalties specifying a payment based on the production from the mine. This was deemed necessary because the landowners started to realize that the coal was a wasting asset and that they should seek some return from this asset rather than from the use of the land on which it was produced. These royalties were initially very high in comparison with the royalty fees which are payable today, although this was partially due to the high rates of inflation which were evident during the period 1540 to 1620. By the late seventeenth century royalty rates were of the order of 5d/t on average although some were as high as 1s/t, around a quarter of the pit-head selling price. In his *Inquiry into the Nature and Causes of the Wealth of Nations* Adam Smith mentions that 'in coal mines a fifth of the gross produce is a very great rent; . . . and it is seldom a rent certain, but depends upon the occasional variations in the produce' (p 272). One reason for these high rents, or royalties, was because of the one-third rents which were paid on agricultural produce at the time. The mention that the one-fifth payment is 'a very great rent' is to emphasize that it is still less than agricultural rents despite the much lower security which mining operations can offer. Smith expanded this by stating that the variations in production 'are so great that, in a country where thirty years' purchase is considered as a moderate price for the property of a landed estate, ten years' purchase is regarded as a good price for that of a coal-mine'.

The middle men who transported the material into London also charged their fees and the retail price of the coal was about 24s 3d for a London chaldron at Westminster in 1700. The 'chaldron' was a unit of weight that was in use in both London and the Tyne. However, the Tyne chaldron seems to have varied in size over time and sometimes makes comparisons difficult. In London this unit of measurement was equivalent to some 28.5 hundredweight or 3192 lb (1.45 tonnes) and the price of coal therefore works out at some

16s 9d/t – a mark-up of about 600%. The high rates of inflation during the period obviously have to take some of the blame for this – especially if the merchants had to tie up capital in the coal whilst it was being transported from the coalfields to London – but much of the increase in price is related to the actual cost of transport.

Nevertheless, the merchants had something of a monopoly on the transport of coal from the coalfields to London and made use of this by making sure that their profits kept pace with the inflation of the period. In 1590 the Lord Mayor of London complained that the price of coal had risen from 4s to 9s a chaldron since the early 1580s. This was when the Grand Lease of Newcastle traders was set up as a monopoly, and in 1600 it was formally incorporated by Royal Charter as the Company of Hostmen. There were originally 24 important partners in the Grand Lease but by 1622 the number of members of the Company of Hostmen had grown to 31, as the profitability of the enterprise became known. But, despite the complaints of high prices and pollution, London remained the main market for coal for house fires until the industrial revolution had started in earnest.

Industrial uses

As the cost of transporting coal in the early 1700s was so great the expanding metal mines started to ship their output to the coalfields rather than the other way around. For instance the very large amount of coal needed to smelt the copper and tin produced from the mines of Devon and Cornwall meant that it was cheaper to take the ore to Swansea for smelting. Here coal was around half the price that would have to be paid in Cornwall, and it led to the start of the metal manufacturing operations in the Welsh valleys. Coal still had to be transported to Devon and Cornwall to power the pumping engines but this was relatively insignificant in comparison with the amount of coal needed in the smelting operations. Bristol also lost out to the cheaper fuel from Wales despite its better position for the end markets which were supplied. Indeed, by 1750 over half of the total copper and lead production of the country was smelted on the South Wales coalfield.

The iron industry also gained in importance as a user of coal during the early part of the eighteenth century. In the latter part of the seventeenth century the first Abraham Darby (there were three) moved his ironworks to Coalbrookdale. He chose this location because the town is situated on the river Severn, so transport was readily available and there was sufficient water power to work the bellows of the blast furnace, and it is in the Shropshire coalfield, in an area where limestone is present. Darby's choice of location was especially fortuitous because the coal produced in the vicinity could be used in a blast furnace for smelting iron, in place of the previously used charcoal. Following the success of his experiments Darby constructed a new ironworks, utilizing his process, which opened in 1709.

The use of coal in industry obviously took off with the industrial revolution but it should not be forgotten that there were many other industries which became dependent on coal for their fuel requirements before the revolution began in earnest. In particular the glass and salt industries burned coal to provide the heat needed to make their products; and in Scotland, during the eighteenth century, salt-pans and coal mines were normally under the same control. Gray also states in his 1649 Chorographia that in 'Warwickshire, in the county of Durham where is many salt-pans, which makes white salt out of salt-water, boyled with coale'. Additionally, other industries needed to be close to the area where the iron and steel which they used was produced. The concentration of the textile industry in the north of England and Wales was initially so that the new mechanical looms could be water-powered. It was only later that looms worked by steam engines powered by coal became important (see Chapter 2). Coal and coke had also been used for a long time in blacksmiths' shops, well before it had been used in refining. For instance Ambrose Crowley, probably the largest ironmonger in Europe at the start of the eighteenth century, situated his nail factory close to his bar iron plant in Sunderland because of the proximity of cheap coal and agricultural produce.

Coal also started to become important to the shipping industry. As already mentioned, the transport of coal in coasters from the north-east coast to its main market in London played a significant, if not the most important, part in the increase in the size of the British fleet. The expansion of the British colonies also started to become important as the eighteenth century progressed. In particular the colonies were prevented from refining pig iron, which they were encouraged to produce in order to reduce British dependence upon Sweden and Russia for supplies, but were forced to ship it to Britain. This was because they would then have been in direct competition with the refineries in Birmingham and Sheffield and these refineries would still have been dependent on outside sources of supply. In order to encourage their production the import of pig iron into Britain was allowed without a duty being charged.

Internal transport by ship or barge also started to increase during the period, although it was not until the industrial revolution had started that construction of canals became a boom industry. In 1724 there were some 1000 miles of navigable rivers, an amount which had doubled over the previous century. For instance Liverpool council stated its desire to improve the waterways in the south Lancashire and Cheshire district and ordered surveys to be taken to help in the task. In particular it wanted to improve transportation between the river Ribble and the Wigan coalfield, and along the Mersey, both south using the river Weaver to get to the Cheshire saltfield at Winsford and up the river Irwell to that coalfield.

Other areas where coal started to be used instead of wood included the manufacture of bricks and in the brewing industry. It was also used in the production of paper and in sugar refining, whilst the expansion of the armaments industry led to an increase in the demand for the production of ever higher quality metals. It is, however, strange to think that despite all of this expansion in the industrial use of coal and its cost of transport away from the place where it was mined, some two-thirds of all coal dug in Britain was still used in home consumption in 1842. However, this was before the boost the great railway expansion gave to the metallurgical industry, covered in Chapter 2.

THE FINANCIAL BACKGROUND

The first coal mines were small scale affairs with little need for capital beyond the work which was put into them by the miners. An early mine at Coundon cost 5s 6d to set up in 1350, including the cost of ropes, scopes and windlass. As the mines themselves were not deep because of the expense of extracting water from below the water table, it did not take long from the time a mine was started until it began producing coal and paying its way. The expansion of industry during the seventeenth and eighteenth centuries required ever larger amounts of coal and the consumers also became dependent upon a secure source of supply. Both of these factors were important in the move of many manufacturers and merchants into the coal mining business, in addition to the need to control the cost of supply.

One of the advantages these early entrepreneurs and capitalists had was a relatively cheap source of capital. Even though there was a relatively high level of inflation during the sixteenth and early seventeenth centuries the Usury Laws kept the cost of capital very low. The rate of interest was lowered from 10% to 8% in 1625 and then further reduced to 6% in 1657 and 5% in 1714. These and subsequent interest rates are shown in Figure 1.4. It should be further mentioned that these were the maximum rates permitted under the Usury Laws and that it was often possible for entrepreneurs to borrow money at more advantageous rates. The high level of inflation and worries about the consequent erosion of capital in the sixteenth and early seventeenth centuries must have led to an increase in the wish to lend.

Nevertheless, the capital markets were still very low key, especially

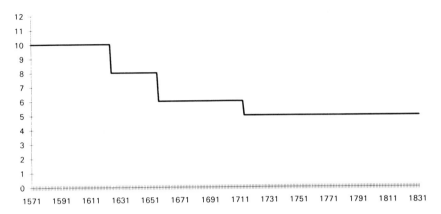

1.4 Maximum permitted interest rates under the usury laws 1571–1832 (%)

following the South Sea Bubble affair in 1720. This halted the development of the limited company with equity capital until, in the nineteenth century, industry became starved of cheap finance and pressed the government to bring in legislation which eased the formation of new companies. Before this only a few companies were set up with the equivalent of limited liability under the aegis of a Royal Charter, and most of the development of the coal mining industry was undertaken by groups of merchants who had access to family capital and who could also borrow money from each other. This is shown in Gray's Chorographia where he mentions 'some Londoners of late, hath disbursed their monies for the reversion of a lease of a colliery, about thirty years to come of the lease'. But he also warns of the risks inherent in coal mining when he carries on to state that 'when they come to crack their nuts they find nothing but the shells; nuts will not keep thirty yeares; there's a swarme of wormes under ground, that will eate up all before their time, they may find meteors, ignis fatuus, instead of a mine'.

This access to capital was important as the industry developed and expanded because of the need to pay for the technological improvements which were being made and which could increase the production of coal from a mine. In particular the development of the steam pumping engine was important in that it enabled the mechanical rather than the manual removal of water from most underground mines accessed by a shaft. In 1698 Thomas Savery devised a steam pumping engine for the removal of water from underground mines in Cornwall. Unfortunately the pump was inefficient in its energy consumption and it was necessary for Thomas Newcomen to make energy saving innovations in 1708 before the pump became more commercially

13

viable. These then started to diffuse throughout the country and by the close of this pre-industrial period there were around 100 pumping engines in the Tyne and Wear area.

THE SOCIAL COST

There can be no discussion of the history of coal mining in Britain without reference to the social cost of the industry. Obviously the early miners had little in the way of protection and rights of employment and the yearly bond by which a miner was tied to a specific colliery was often the subject of a dispute. The other main problem was the legislation of the period with one particular Act in 1610 stating that any able bodied person who even threatened to run away from his or her parish could be sent to a house of correction and be treated like a vagabond.

The other restriction of the yearly bond was that it would only last for a year less one day. This was because of the Poor Law legislation which meant that a parish had to look after any inhabitant who had spent more than one year within its boundaries. If a miner had to travel to a different parish to work there was often much disaffection between the new arrivals and the original inhabitants. This was particularly the case with small villages far from large scale habitation which could be overwhelmed by an influx of miners to open up a new colliery. In 1606 the inhabitants of a Shropshire village complained to their landlord that the new colliery had brought with it 'a number of lewd persons, the scum and dregs of many [counties] from whence they have been driven' (*Reformation to Industrial Revolution* Christopher Hill p 59).

As already mentioned, the seventeenth and eighteenth centuries saw an increasing amount of industrial specialization as miners started to concentrate fully on mining with little time spent on the farms of their landlords except, perhaps, at harvest time. The sparse population in many rural areas also failed to help matters as the increasing need for manpower meant that local food production could not be maintained if a mine was to be opened up. It was therefore necessary to import labour into new coalfields and to ignore the resentment shown by the indigenous population. One of the areas particularly deficient in inhabitants was Wales and the Society of the Mines Royal was given the right to conscript labour in the country in 1625.

The power of the local lord was often sufficient to put down any complaint during the pre-industrial age. This was partly because of the relatively small number of employees working in any one mine and partly because of the nature of society at that time. Indeed, the feudal nature of the coal mining industry in Scotland continued into the nineteenth century with the employees being included in any purchase agreement over a mine. It was only during the late eighteenth and early nineteenth centuries that the increasing size of the workforce meant that it could start to organize itself to greater effect. It was also the case that the increasing movement of people to towns and cities meant that the feudal system in Scotland started to break down and that the control which could have been exercised in the past became less effective. The increasing wealth of merchants and the power which this conferred on them also meant that the authority of a local landowner was reduced. As a result the merchants came to usurp the previous system of feudal control and often purchased property directly, whether for mining or agricultural purposes or both.

Linked in with both the capital cost of mining the coal and the social cost of its extraction is the ownership of the resource. This has been touched on briefly with the mention of the leases and royalties which producers had to pay to the freeholder of a property. There were always disputes about the ownership of a particular resource and in some instances copyholders were given an allowance of free or cheap coal in lieu of 'firebote', the right to take coal for their own consumption. In some instances the rights of the copyholders were maintained until the nationalization of the mineral rights in 1942 when they received compensation as if they had been freeholders (see Chapter 4).

When a landowner wanted to lease mining rights in a particular area the miners could be charged both rents and royalties. The leases which were offered were normally for a period of 21 years at most and were sometimes only for a shorter period. This meant that the first mines were relatively short life operations and that large amounts of capital could not be recouped unless the owner had the freehold rights to the property and could be guaranteed long term tenure. Nevertheless, in these early days the small scale of the operations and the control which many miners seem to have been able to exercise over their own income as a result of being paid on the basis of production, is in marked contrast with the extremes of control exercised over large scale operations in the late eighteenth and nineteenth centuries. This is covered in Chapter 2.

Chapter

2

The first coal age (1765–1913)

INCREASING DEMAND

Up until the middle of the eighteenth century the consumption of coal was restricted by the location of the resource and the availability of economic transport. As transport by sea and canal expanded and reduced the delivered price of coal its increasing use became more dependent upon the inventive skills of the population. The discovery that coke could be used in place of charcoal to smelt iron ore into pig iron was a useful addition to its capabilities. However, a great bottleneck still existed with the refining of the pig iron into iron and steel, and the lack of iron was the major constraint on the advance of the industrial revolution. This was not overcome until 1784 when Henry Cort developed a puddling process which, like Abraham Darby's discovery 75 years before, enabled coal to be used in place of charcoal. Following this discovery all new sites for the production of iron relied on their proximity to deposits of coal, iron ore and limestone. British output of pig iron and coal is shown in Figure 2.1. This clearly demonstrates the link between an increase in coal supply and demand and the rise in the production of iron.

Smelting iron was not the only use which increased the consumption of

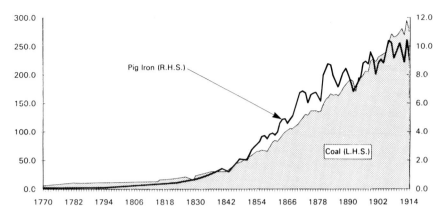

2.1 British coal and pig iron output 1770–1914 (million tonnes)
Source: Mitchell & Deane.

coal during the period. Indeed, the great industrial advance of the country across all sectors of the economy was important, as was Britain's position as the lynchpin of an expanding Empire. The growth of world trade and Britain's central role in that trade meant that the colonies came to depend on Britain for most of their imports. This was particularly true following the 1805 defeat of the French at Trafalgar, which confirmed Britain as the most powerful seafaring nation until the Germans started to build up their naval capability after 1898. And the two-way trade in finished goods and raw materials was enhanced because legislation meant that many of the goods required by the colonists had to be made in the UK rather than produced locally. This meant an increase in British demand, not just for the iron necessary to make all types of tools and implements, but also most importantly for textiles.

Whilst textiles were initially made in a cottage industry the inventions of larger, mechanical, looms meant that individuals could no longer compete with larger organizations. This led to the start of the large scale milling industry in the north. The location here, especially in the Lancashire coalfield, seems to have been as much to do with the proximity of the Port of Liverpool (which handled the imports of raw cotton and then the re-exports of finished materials to the colonies) as it was related to the need, initially, for water power to work the looms. It was only by accident that the mills were constructed close to the northern coalfields as the first steam-powered mill was not built until 1785 at Papplewick in Nottinghamshire. Indeed, the cost of power was not a significant amount of the total cost of operating a mill and, if required, mills could still have been operated using water power instead of steam. Indeed, as all of the early milling machines were made of wood they were unable to take the strain of steam power. This advance had to wait for looms to be made from the much more resilient iron and steel and was only achieved because of the need for advances in metallurgy and manufacturing techniques necessary to increase the fighting strength of the army.

Between 1785 and 1838 some 80% of the mills in the country converted to run on steam power or closed down if they were unable to obtain coal and were seemingly inefficient. The move to Lancashire was just as staggering as by the same year three-quarters of the 1600 mills in the country were situated in that county. The move to steam continued apace and by 1850 around 90% of the mills were using steam power. This level was effectively a maximum as the use of water continued for many years in local districts and the mills were later converted to electricity. The actual number of mills, however, only tells half the story as the growth of the industrial base, and expanding exports, called for larger and more efficient mills to be built. The numbers of looms within the industry showed a phenomenal rise from 1813, when there were only an estimated 2400 power looms in the country, compared with the much greater 224 000 which were in use in 1850.

The dispersal of the use of coal to fire steam engines to work textile factories is an example of the time lag which occurred between the invention and the widespread application of coal and steam power in many industries. Whilst the use of coke to smelt iron was first shown to be feasible in 1709 it was not until the 1760s (and the start of the industrial revolution) that the technique became widespread. Before this discovery it was actually the case in the late seventeenth and early eighteenth centuries that the production of

Table 2.1 The use of different types of steam pumping and winding engines in coal mines during the late eighteenth century

Year	Area	Type	Use	Number
1750	West Riding	Atmospheric	Pumping	8
	Derbyshire	Atmospheric	Pumping	3
1769	Newcastle upon Tyne	Newcomen	Pumping	100
1798	West Riding	Atmospheric	Winding	1
	West Riding	Boulton & Watt	Pumping	2
1800	West Riding	Atmospheric	Pumping	53
	Derbyshire	Atmospheric	Pumping	14
	Derbyshire	Atmospheric	Winding	1
	Derbyshire	Boulton & Watt	Pumping	1
	Scotland			77

iron was falling! Problems with the spread of information and inventions also affected Watt's design of the rotary steam engine, which he perfected in the 1780s. However, it was not until 1800 that his patent expired and use of the invention diffused throughout the country. Table 2.1 shows the spreading use of different types of steam pumping and winding engines in coal mines during the late eighteenth century.

PRICES AND TRANSPORT

The use of coal across the country had another problem – transport. As discussed in Chapter 1, the British road network was poorly maintained and often impassable during spells of bad weather. This meant that internal trade was often difficult and inefficient and that despite the vast expansion in navigable rivers there was still little easy access to many parts of the country. Where canals had been started they were only relatively modest in scope and did not have a great effect on the movement of produce around the country. However, the last 30 or so years of the eighteenth century saw a massive move to canal construction with the result that many prices fell, sometimes drastically, as the availability of produce increased. Indeed, this fall in prices is a prime example of the thesis of price being linked to supply and demand.

The most famous canal of the period was that constructed for the second Duke of Bridgewater between 1759 and 1761, in furtherance of a project started by his father. The canal was built by James Brindley and included an aqueduct and 42 miles of underground waterway to take the Duke's coal from his colliery at Worsley almost to Manchester. The last stretch of the canal into Manchester was completed in 1763 and was helped by the low level of interest rates at the time. The construction of the canal at a price of some 10 000 guineas a mile led to a halving in the price of coal in Manchester and to a massive increase in the number of other canals proposed. Indeed, like motorways which attract traffic, this reduction in the delivered price of coal stimulated demand to such an extent that the first canals provided a high initial return on investment. Such was the increase in the demand for coal from the Duke's mine that he is reported to have said that 'a navigation should have coals at the heel of it' (Mathias, *The First Industrial Nation*, p 112). And in 1785 J. Phillips' book *A Treatise on Inland Navigation* stated that the canal had proved successful 'by drawing a mine of wealth from the bowels of a mountain which had hitherto been totally useless and of no value' (p 85).

The success of the Bridgewater canal and the massive boom in canal

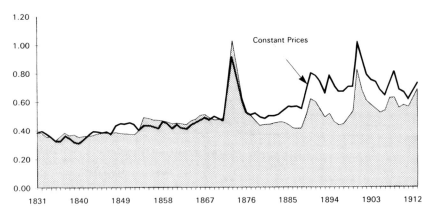

2.2 Coal prices FOB export 1831–1913 (£/tonne)
Source: Mitchell & Deane.

construction across the country can be shown by the large number of private bills sponsored through Parliament to authorize the building of the new waterways. Of the 165 Canal Acts which were passed through Parliament between 1758 and 1802 there were 90 which expected coal to be the major commodity carried. The main link from the Midlands to London was started in 1793 and completed in 1805 when the French privateers were causing trouble along the coastal routes. And a total of 160 miles in the vicinity of Birmingham gave a great impetus to the construction of collieries in the Midlands coalfield as the canals gave access to the sea as well as to London. Between 1724 and 1815 the increase in the availability of water transport had been phenomenal, but was still less than the advances which the railways were yet to provide. The length of navigable rivers had doubled to 2000 miles and some 2200 miles of canal network had also been constructed.

All of these advances together with the advances made in underground mining techniques and the number of mines supplying coal led to a decrease in price. This fall in price is shown if Figure 2.2 is compared with Figure 1.3 which details the price of coal delivered to Westminster School in London. Figure 2.2 also includes a real price index to show how the deflation in the latter part of the nineteenth century led to a real rise in the price of coal.

THE RAILWAYS

The invention of the rotary steam engine and the age of the train led to a massive increase in the demand for coal. Not only was it

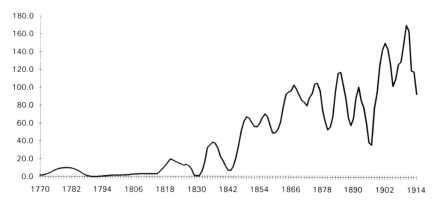

2.3 Growth in UK coal production, rolling five year periods 1770–1914 (million tonnes)

obviously required for the powering of the engines themselves but also it was needed to make the iron both for the trains and for the tracks they ran on. The great age of the railways really started in the 1840s and continued for much of the rest of the century, helping to spur the development of the coal mining industry at the same time. Indeed, between 1841 and 1901 the mining areas attracted around 500 000 people from rural parts of England and Wales and by as early as 1851 there were 216 000 men and 3000 women employed as coal miners throughout the country.

The total output of coal between 1760 and 1840 had increased from 6 mt to around 30 mt but the increase in production as the pace of progress quickened during the rest of the century is considerably greater (Figure 2.3). Most of this increase was for use in industry rather than for domestic consumption which had been the main area of demand until the middle of the nineteenth century. This was directly related to the use of coal in making iron, as 1 tonne of iron required 4 tonnes of coal. With the output of iron quadrupling the output of coal also had to increase.

Whilst the first important public railway line with a moving steam-powered engine, the Stockton and Darlington, was opened on 27 September 1825, the initial work on steam engines had been going on since the turn of the century. This was encouraged by the need to transport coal from the pit-head to roads or rivers where it could then be shipped to other parts of the country. This started during the seventeenth century with large collieries laying wood on the ground to ease the movement of wagons to the local river. In the early eighteenth century the greater availability of iron meant that the wood started to be replaced by iron plates, particularly at bends where the wear was greatest. In 1767 the first track was laid by Richard Reynolds from Coalbrookdale to the river Severn. It used rails with flanges to keep the wagons in place but as this was found to be inefficient at keeping the wagons on the rails the flanges were transferred from the track to the wheel following the advice of John Smeaton in 1789.

Although the first public railway line in the country was opened on 26 July 1803 between Wandsworth and Croydon, the Stockton and Darlington was the most important in terms of its effect on the movement of materials. Like the development of the canals some 40 years earlier this railway's prime motive was to transport coal in order to break the effective monopoly on the production and distribution of coal held by the colliery owners of the Tyne. The construction of the railway was intended to open up the country to the south so that new mines could be dug and the monopoly broken. It was the idea of a group of three Quaker businessmen, Pease, Richardson and Backhouse, and was made possible because of the invention of a movable engine rather than one which was fixed in position and wound wagons using a cable.

The first successful experiments were those of Richard Trevithick who, in 1803, developed a steam carriage and then a locomotive to run on rails which hauled a load of 10 tons of iron, 70 men and five additional wagons over 9.5 miles on 22 February 1804. Nevertheless, it still required the demonstration by William Hedley in 1812 that two iron contacts could create sufficient friction to prevent the wheels from slipping on the track and that a rack system was therefore unnecessary except on steep ascents. The power of the engine also had to be increased so that it could pull more than its own weight and this was achieved by George Stephenson who was able to improve engine efficiency by increasing the draught of the fire box. He first built locomotives for the Killingworth Colliery in 1814 and served as the engineer for the Stockton and Darlington railway. This experience must have helped him to win the Rainhill competition on the Liverpool and Manchester line between 6 and 14 October 1829, which showed that steam engines could offer faster travel than other alternatives.

As with the canals, the opening of new railway lines reduced the price of coal in the areas they served and this helped to increase demand for the product. By the end of the 1840s a skeleton network of some 5000 miles of track had been completed, whilst by 1886 the final total of 16 700 miles of track had been laid. Overall, however, there were periods during which the expansion of the mining industry could not keep pace with the development of the railways and the price of coal was forced up. This was often only a local effect in areas of iron and steel production as manufacturers needed to purchase coal in order to maintain and increase output. It must also be related to the integration of many coal and iron companies which would supply their own plants first before selling the remainder of their production on the open market. If there was insufficient coal to meet demand then the price would be forced up because of the need to transport it from collieries further afield.

By 1850 the amount of coal used on the railways was only around 1 mt, a relatively small figure representing only some 2% of total production. It was, however, the knock-on effects which were significantly greater and which led to the rise in consumption. The technological advances of the period also played a part in the increasing demand for coal by the railway industry and of particular importance were the inventions of the Bessemer Converter (1856) and the Siemens-Martin Open Hearth (1866) processes for the production of steel. Both of these inventions meant that steel could be produced much more effectively and cheaply than previously and that it was economic to use it in place of iron on railway tracks. This was desirable because steel was more resistant to erosion and corrosion than iron and replacement of the track did not have to be carried out so frequently. These advances led to a fourfold increase in the production of steel between 1875 and 1895 as it replaced iron rails and also found applications as mild steel plates and girders in ships. This increased the amount of steel-hulled shipping and meant that the construction of steam ships was easier than before. In 1851, despite the long existence of the steam engine by this stage, there was only some 188 000 t of steam shipping out of a total of some 3.9 mt, and the majority of this was still in wooden ships. By the close of this period oil also started to be imported in moderate quantities and in 1913 Winston Churchill, then First Lord of the Admiralty, announced that the Royal Navy would convert its ships to oil firing. Arguments over the security of supply, and hence the security of the nation, emerged and in order to allay some of these fears the Admiralty purchased 51% of the share capital of the Anglo-Persian Oil Company (now British Petroleum) for £2 million in 1914.

A DANGEROUS BUSINESS

The slow advance of technology meant that mining during the nineteenth century was a very dangerous occupation. It remains hazardous today but the advances in technology make it considerably less so even than in the recent past. One of the reasons for the increasing danger during the nineteenth century is related to one of the basic facts of mining. The first tonne is the easiest and safest to extract and the risks increase, almost at a compound rate, thereafter. This occurs because the first tonne is removed from the bottom of a shaft, or the start of a decline, whereas the last tonne will be taken from the furthest point away from the shaft bottom with all of the ventilation and support problems that entails.

In the 12 years to 1867 there were a reported 12 590 people killed in colliery accidents in Britain. In 1867 this represented an average rate of 357 for every 100 000 employees. The rates have fallen considerably since then but the difficulty now is that whilst there may be no accidents in any one year the problems still exist, especially because accidents often involve a large number of people. This preoccupation with the value of life in the industry led to the formation of some of the most militant unions to represent the miners in their claims for better pay for the work which they undertook in such frightening conditions.

For many years the miners had been poorly paid and their pay had not increased in line with the increase in the price of the commodity they were producing. One of the reasons for this was the poor productivity of the industry which continually needed to increase production and sought to achieve this by employing more miners rather than by improving mining techniques. Indeed, by the end of this period, in 1913, the British coal mines were much less productive in terms of annual output per employee than those of either Germany or the USA, despite Britain having a much greater overall level of output. This low level of productivity and the problems associated with the Great Depression of 1873–96 led to calls for a coal combine or cartel to attempt to maintain prices and keep the industry profitable.

PRODUCTION TECHNIQUES

After the invention of the pumping engine the large number of small mines exploiting seams which were only 18 inches thick delayed the introduction of new technology. As a result the mechanization, and hence productivity, of the operations lagged behind other countries where thicker seams enabled more profitable extraction. It is also true that many of the UK's easily accessible thick seams had been worked in earlier times and that only the thinner near surface ones were left. The deeper seams which are now being mined are also relatively thick and can be extracted using mechanical techniques, but they were too deep to be mined during the last century.

There was only one major advance in the production of coal during the entire period of the industrial revolution – and some said at the time that it should not have been considered as an advance at all. This was the invention of the safety lamp. Most of the credit for the invention went to Sir Humphrey Davy but his was only one of three lamps invented between 1813 and 1815 and it seems to have worked by default as he did not understand the real reasons for the success of the lamp.

The need for safe illumination underground was well recognized in the industry and the sinking of deeper mines in order to increase output necessitated the development of a new method. This was because the deeper the mine the poorer and more restricted ventilation became. In 1812 an explosion in the Brandling Main (or Felling) Colliery, near Durham, led to the formation of the Sunderland Society for the Prevention of Accidents in Coalmines. In 1815 the Society appealed to Sir Humphrey Davy to investigate methods of improving the safe illumination of the mines and this led to his development of the safety lamp.

The other lamps were produced by Dr Clanny and George Stephenson and worked on a similar principle, restricting the amount and temperature of the air passing over the flame and thereby reducing the risk of an explosion which could have been caused by a higher temperature ignition. The problem that the miners faced as a result of these advances was that they were able to extract coal from areas with poorer ventilation. As a result the likelihood of asphyxiation from excessive concentrations of carbon dioxide (blackdamp or chokedamp) and the poisonous carbon monoxide (afterdamp) rather than the explosive methane (firedamp) increased. These increased dangers led John Buddle to complain to the 1829 House of Lords Committee into the mines that 'we are working mines from having the advantage of the safety lamp, that we could not have possibly worked without it'. And, although it was not so clearly recognized at the time, it would have enhanced the likelihood of the miners contracting pneumoconiosis and other breathing disorders.

This is not to say that there were no important innovations in the industry except for the development of the safety lamp. Indeed, in 1777, long before it was used above ground, the cast iron rail was being employed by Joseph Curr to aid underground transport. A wheeled corf, or wagon, was also invented by him to run on the rails and had the advantage of being able to be hauled straight up the shaft with no need to move the coal from one container to another, again reducing costs. These corfs were also hauled along the tracks using pit ponies rather than the women and boys who had previously been employed for the task. This meant that the miners could keep more of their money without the need to pay others to haul their coal out of the mine. However, it meant that there was a large number of unemployed women and girls in the mining towns and villages and this labour was a further attraction to the area for the textile millers who were setting up at the time.

Ventilation remained a problem for many years and is still one of the most difficult and dangerous areas in mining. The main reason for this is the unpredictability of rushes of methane out of the coal seams which can then be ignited to cause the dangerous explosions which have claimed so many lives. The accident at Haswell Colliery in 1844 is a fearful example of the problems of insufficient ventilation. The mine suffered an explosion as a result of excess coal dust in the air underground igniting. This dust was not fully burned and meant that the air had a large amount of afterdamp, or carbon monoxide, in it following the explosion and it was this which killed most of the 95 miners who died, not the direct effect of the blast.

Another failing of the ventilation system was brought to light by the accident at Hartley Colliery in Durham in 1862. The beam of the pumping engine at the mine broke and crashed down the shaft. Under most circumstances this might not have been a serious problem but, unfortunately, the mine only had a single shaft with a brattice separation in order to separate the air flows through the mine. One side of the shaft was for the clean air going in and the other for the used air coming out. The iron beam crashed through the brattice and disrupted the air supply to the miners who were all unharmed by the fall of the beam. However, as the oxygen at the bottom of the mine was used up the miners slowly suffocated and all 204 in the pit at the time of the accident died. This led to the requirement that coal mines had two shafts in order to attempt to ensure that there was a sufficient flow of air through the mine at all times.

Brattices had earlier been invented to lead the flow of air through the underground workings of a mine and prevent it from dissipating without having any beneficial effects. This idea was developed by Carlisle Spedding of Whitehaven in the 1750s and was enhanced by John Buddle who invented a system involving three shafts and a more elaborate method of 'coursing' or leading the air through the workings. Sometimes if the two shafts were at the same topographical level the air had to be helped to start it moving through a mine. This was often achieved by lighting a fire at the bottom of the shaft which the engineer wished the air to escape from and the subsequent loss of pressure in the pit sucked air in through the other shaft. Fans were not powerful enough to be useful for the ventilation of mines until well into this century.

THE START OF UNIONISM

The poor working conditions of the miners have already been touched on in this chapter. Nevertheless, it is important to realize that many of the problems encountered were beyond the control of the workers at the time. The British population was increasing at a much faster rate than it had in earlier centuries and this led to a greater pressure on the workforce to find employment in order to be able to afford to live. This meant that there was often little option but to accept the conditions laid down by the management of a mine or other operation during the early part of the period although the negotiation of the yearly bond and introduction of a leaving certificate led to a strike of 4000 miners in the Durham area in 1765. Such a show of militancy was not often repeated due to the low legal standing of the workforce which can be seen from the civil offence with which employers were charged for breach of employment contract rather than the criminal offence with which employees would have been charged for a similar offence.

As time progressed the widening of the voting register helped to introduce a fairer spread of views in Parliament with a succession of Commissions, inquiries and legislation during the century. As far as the mining community was concerned the first whisperings of unionism had started in the years following the defeat of the French at Waterloo. Moreover, as the scale of industrial operations became larger it became necessary to treat the workforce on a collective basis. It followed that the repeal of the 1799 and 1800 Combination Acts was required so that the formation of employee groups, which could engage in collective bargaining, was possible. These Acts were repealed in 1824 and 1825 and legalized the formation of the first proper mining unions. However, the financial shortages during periods of depression, as a result of strikes or in the extreme event when union organizers absconded with the money, meant that none of these early unions was long-lived.

The first recorded union was variously called The Coal Miners' Friendly Society or the Pitmen's Union of Tyne and Wear and was organized under the leadership of Tommy Hepburn in 1830. Early in 1831 the miners denounced the yearly bond together with the wages and tied cottages attached to it, colliery 'tommy-shops', fines inflicted by viewers, and the long hours which 12 year old boys were forced to work underground. The yearly bond, in fact, was not finally abolished in the north east of England until 1872 and had long been a cause of contention, being behind the strikes in 1765. Colliery 'tommy-shops' caused problems because the mine owners often paid their employees in tokens which had to be spent in their own shops where inflated prices were charged. A worse form of this was when the employees were paid in public houses.

As far as the fines were concerned the miners were worried that the viewers, who were employed by the owners to ensure that the miners were filling their skips with coal rather than waste, were imposing fines that were too large and that they were not taking full account of the amount of coal that had been hewed. This again resulted in a long campaign leading to the creation, in 1860, of the independent position of 'checkweighman' whose job was to ensure that miners were paid for the amount of coal that they had produced. However, it was not until the 1872 Coal Mines Act provided that the miners should be paid according to the amount of coal produced and 1887 when the checkweighman was given full authority that the situation was resolved. The final disagreement over the amount of work required by young boys was eased by the 1842 Mines Act (see below).

The union organized a strike in the Tyne and Wear coal field following the expiry of the yearly bond on 5 April 1831. Before this they had obtained an agreement that allowed them to spend money where they wanted to and that the number of hours worked was to be limited to a maximum of twelve. By mid-June the owners had capitulated on the problems surrounding the yearly bond and had increased wages in order to cover some of the problems of fines. As a result the miners returned to work but when the yearly bond negotiations were underway the following year the owners refused to employ trade union members and brought in workers from the local metal mines to break the strike and force the members to renounce the union, which effectively collapsed.

In 1841 The Miners' Association of Great Britain and Ireland was formed at Wakefield in Yorkshire and by March 1844 it had enrolled 52 927 miners. This was around 25% of the total number of miners in the country at the time. The following month the union organized an unsuccessful strike which collapsed after 20 weeks. Although there was a brief rally in 1850/51 the union had finally disintegrated by 1855. Following this the South Yorkshire Miners' Association was formed in 1858 and in 1863 the Miners' National Union (MNU) was organized by Alexander Macdonald (see below). The MNU became engaged in many disputes during the years 1867–68 and 1871–73. The second series of disputes was particularly successful and came after the formation of the more militant Amalgamated Association of Miners by Thomas Halliday which covered the South Wales, Lancashire and Staffordshire areas. Two years later, in 1875, the two unions merged and lasted for a further six years until Macdonald's death in 1881.

However, by the end of the 1870s the depression was forcing the miners to accept conditions that negated many of the advances that had been achieved over the previous decade. But, in 1888, as the economy started to recover the miners' longest lasting union, The Miners' Federation of Great Britain (MFGB) was created in the inland areas of Lancashire and Yorkshire. By 1892 it had enrolled over 300 000 members (almost 44% of the total number of miners) and was joined by the Northumberland and Durham miners in 1907. The union lasted for 56 years until the formation of the National Union of Mineworkers (NUM) in 1945.

In a completely different view from those expressed today the union is reported to have argued on 30 June 1893 that 'pits which could not afford to pay a living wage should close'. This statement was made in response to a request from the mine owners for a 25% reduction in wages and resulted in a strike. By the end of September the MFGB allowed the miners to return to work if their employers agreed to continue to pay wages at their pre-strike level. As a result just over 87 000 men returned to work and enabled the production of coal, which was then selling at high prices and leading to high profits for those mines still in operation. Those pits which remained closed with their 228 485 employees locked out were therefore pressured into a settlement which was not forthcoming until eventually the Prime Minister, Gladstone, arranged for the two sides to meet with Lord Roseberry, the Foreign Secretary, as arbitrator. They agreed to set up a conciliation board which resolved the dispute, although it continued to be active until the end of the war in 1918.

However, it was not until 1912 that the MFGB was, after the 1907 joining of the Northumberland and Durham miners, able to agree with the 1911 Unofficial Reform Committee of South Wales miners, and call the first truly national miners' strike. This strike started at the end of February 1912 and lasted until early April with the government agreeing to the introduction of a national minimum wage which was enacted in the Miners' Minimum Wage Bill later that year.

The spread of political thought also moved into the mining communities during the nineteenth century, although there appears to have been little influence from the Chartist movement in the 1840s. This may have been part of the reason for the delay in the miners getting the right to vote until the Reform Bill of 1884. This followed from the previous Bills of 1867 (when working men in towns were enfranchised) and the widening of the franchise and redistribution of seats which was given in 1832. Despite this the first miners to be elected to Parliament in 1874 (i.e. ten years before the miners were given the vote) were Alexander Macdonald and Thomas Burt, both of whom sat as Liberals.

Nevertheless, one of the most significant inquiries into the state of employment in the British coal mines had already been undertaken, and acted on, some 30 years before. This was the First Report of the Commissioners on the Employment of Children which covered their employment in coal mines, and was published in 1842. The Commissioners set themselves to look at 14 separate aspects of children's employment including their ages and the hours that they were expected to work. One of the most revealing details was the very young age of many of the children who worked in the pits, often with very little supervision and left for long periods on their own in the dark. 'The lowness of the roof or the thinness of the bed of coal . . . is no doubt the cause

Table 2.2 The numbers of women and children employed in the mines per thousand
adult males

County	Adult females	Children aged 13 to 18		Children under 13	
		Males	Females	Males	Females
Leicestershire	—	227	—	180	—
Derbyshire	—	240	—	167	—
Yorkshire	22	352	36	246	41
Lancashire	86	352	79	195	27
South Durham	—	226	—	184	—
Northumberland & N Durham	—	266	—	186	—
Midlothian	333	307	184	131	52
East Lothian	338	332	296	164	103
West Lothian	192	289	154	180	109
Stirlingshire	228	283	129	184	107
Clackmannanshire	202	246	213	142	87
Fifeshire	184	243	109	100	34
W. Scotland	—	223	—	99	—
Monmouthshire	—	302	—	154	—
Glamorganshire	19	239	19	157	12
Pembrokeshire	424	366	119	196	19

of employing boys instead of horses or asses' (to move the coal) was one of the
conclusions of the report which found children as young as six or seven at
work. The publication of the report led directly to the Mines Act of 1842
which forbade the employment of boys under 10 and all females with a penalty
of £5–10 for each offence. The numbers of women and children employed per
thousand adult males (of whom there were around 200 000 at the time) is
shown in Table 2.2.

SUNRISE INDUSTRIES

That the incomplete combustion of coal, both from
underground explosions and from the production of coke, led to the manufacture of a flammable gas had been known for some time. William Murdoch,
Boulton and Watt all carried out experiments at the Soho works in Birmingham
towards the end of the eighteenth century and Murdoch showed that it was

feasible to use it for lighting in 1792. It was, however, not until the start of the nineteenth century that the use and piping of gas was recognized and the first public gasworks was authorized by an Act of Parliament in 1810. This industry further increased demand for coal and for the iron required to make the pipes which had to be laid in a network from the gas stations. The development of electricity was also an important factor in the demand for coal in the latter part of the nineteenth century. The first public generators were built in the 1880s and commercial generation started in Godalming, the first town to be illuminated by electricity, in 1888.

Industrial applications of coal remained limited for the majority of the period, and in many respects still do today. The use of coal as a source of heat for the manufacture of bricks and beer was acknowledged before the industrial revolution began. Coke manufacture had also begun with the need for coke to fire the pig iron manufacturing process invented by Abraham Darby at Coalbrookdale. The use of coal in the chemical industry, even today, remains relatively limited despite the advances made by Sasol in South Africa. The first real investigations, however, were carried out in Scotland by the ninth Earl of Dundonald, Alexander Cochrane. He experimented with the coal on his estate and proved it as a potential source of tar and varnish. This success prompted him to set up his own manufacturing plant at Culross in 1792 but a lack of capital and problems with the Admiralty led to a series of losses. It was then up to John Loudon Macadam to develop the idea from which he reaped the rewards, and left his name to posterity.

FINANCE AND CAPITAL

One further constraint on the mechanization of the British coal mines was the lack of available capital in the industry. In the London equity market many of the mining capital raisings concerned gold mining ventures in Australia and South Africa during most of the nineteenth and early twentieth centuries. Coal mining was seen as the poor man's equivalent and the poor profitability record of the industry did not normally make investors rush for new issues of shares. Therefore, despite the extension of limited liability to all registered companies under an Act in 1856, there was little rush to take advantage of this as a new source of capital for the industry. Despite this, many of the existing coal operations transferred themselves to limited liability companies in view of the protection this then offered to shareholders. This was particularly important because a profitable mine could be brought down by the bankruptcy of a shareholder who also held a stake in another mine which had been forced into liquidation by continuing losses. It was intended that the removal of personal liability would prevent this from happening.

It was also the case that the great railway boom of the 1840s attracted much of the capital available for investment in industry. By December 1845 there were 260 different railway shares (including preference and debenture stock rather than just basic equity capital) quoted on the London Stock Exchange and in 1846 a total of 273 railway bills received Royal Assent. These new companies went straight to the equity market for subscriptions whilst the coal mining concerns, which had been in existence for long periods of time, just took advantage of the protection offered by limited liability status. This was particularly following the 1856 Act when limited liability was extended to all companies which had registered under the 1844 Registration Act. In the 1850s and 1860s many private concerns in the coal, engineering and iron industries converted into public companies.

Chapter

3

Wasted lives, wasting assets (1914–1945)

THE END OF AN ERA

In 1913 the British coal mining industry produced more coal than at any time before or since. The start of World War I marks the start of the decline which, even today, seems to be without end. One of the main problems faced by the coal industry during the years of the industrial revolution was that it was always a demand pull industry. In many types of organization this is accommodated in fits and starts by building new factories which can handle the excess demand and provide some scope for further expansion in the future. However, the mining industry of the nineteenth century could not fit into this mould. This was because by this time the industry had become such a conglomeration of small and large companies with small and large mines that any additional mine had little effect on the supply/demand equation.

As it was always the case that more coal was needed, mines were encouraged to use unsafe mining practices and some companies were set up to extract coal from seams which would not be economic under more normal circumstances. Nevertheless, there were still periods of coal famine in various parts of the country which further encouraged expansion within individual mines. This expansion was often achieved through opening up as many new faces within each mine as possible rather than through the comprehensive planning of a new operation employing more productive techniques. The start of World War I, which saw the enlistment of 250 000 out of a total of 1 133 700 miners, also led to a vast fall in demand and a consequent decline in the output of the industry. The need to go out and sell coal, rather than being faced by a succession of eager purchasers, was a similar shock, and one which has hit much of British industry.

The other problem which surfaced with the start of World War I was the beginning of government control and intervention in the industry on a scale which had not been seen before. In the past this intervention was instigated by the miners and their unions who brought their poor working conditions and low levels of remuneration to the notice of a number of Parliamentary Commissions. The national dependence on coal as a source of fuel during the war years meant that the fuel came to be seen as a national asset and that a co-ordinated production programme was a necessity. Whilst retaining private ownership, the control of the mines passed to the government and this divorce of ownership and control must have had some influence on the poor rate of technological advance in the industry – especially because of the inherent conflict between the need for a long term production strategy and the maximization of short term output necessary to help with the war effort. The dichotomy of ownership also exacerbated the rift between the owners and the unions.

DISENCHANTMENT AND DISSOLUTION

As the nineteenth and early twentieth centuries progressed, the structure of the mining industry was seen by many to be an uncompetitive jumble of companies and mines with no formal strategy or operating plan. The easiest and most productive mines in the country were being exploited to the long term detriment of output as the areas furthest away from a particular shaft were abandoned on economic grounds despite still containing workable coal. The abandonment of a mine in such a condition would mean that the coal was lost for ever as the costs of reopening the shaft and digging new tunnels out to the original faces would be so high that the potential earnings from the coal would not provide a positive return on the investment.

It was at this time that nationalization was first proposed in the attempt to guarantee the long term security of the country's coal supplies. Before this talk of national or state ownership was started by the unions which wanted to remove capitalist ownership of the pits. Their original arguments in favour of nationalization gained acceptance because of the spread of socialist and Marxist thought. But the productivity benefits which could be achieved by rationalization (merging two small mines into one big one) also began to encourage cross-party support for a restructuring of the industry. This restructuring, many believed, could only be achieved through nationalization (see Chapter 4) and many Royal Commissions were set up during the inter-war years to report on the structure and prospects of the industry.

Probably the most significant Royal Commission to look at the coal industry in the UK was that of Judge John Sankey in 1919. The first draft of its report commented that nationalization should be recommended because 'coal is our principal national asset, and as it is a wasting asset, it is in the interests of the State that it should be won and used to the best advantage'. However, the final draft was never approved because the Commissioners effectively split along party or interest lines, one Commissioner even preparing and submitting his own personal report because he could not agree with any other members of the Commission. It was therefore impossible to obtain a consensus, and talk of nationalization remained a contentious issue. The report was, however, prepared when the mines were still under state control as a result of the need for the government to manage the mining industry during the war. It was not until March 1921 that they were returned to private control (just before the 1920 pay agreement ran out) and a series of wage cuts around the country led to a lockout which started on 1 April 1921. The railwaymen and transport workers, who were to play such a devastating role in the General Strike, went back on their words to join the miners' strike on 15 April, a day termed Black Friday by the miners.

The one difficulty which the mining industry faced at the time was that the poor working conditions and the dual effects of the war and lack of investment meant that the workforce was becoming disenchanted with the management. One of the reasons for this lack of investment must have been the relatively small number of companies of sufficient size to expand and mechanize the industry. Indeed, by the mid-1930s there were still over 2000 mines in Britain. This was often put forward as one of the reasons for nationalization as it would have led to an automatic consolidation of the industry which, it was expected, would then lead to output and productivity gains. Nevertheless, some of the problems must have been caused by the absence of long term planning as a result of the need to maximize output during the war. That state control was continued for some two and a half years after the cessation of hostilities cannot have helped the situation.

The start of the British Labour movement was led by the mineworkers who had set up unions in the nineteenth century to fight the awful conditions in which they were forced to work. These unions often had a chequered history, normally lasting for no longer than a particular strike over a particular grievance. It is also true that only rarely did they have national support because of the variety of working conditions in different parts of the country. These conditions were related to different working practices which had developed over the years of operation of the mines and often led to miners from separate areas arguing over terminology when they were agreeing in practice. This lack of co-ordination and compatibility may have been one of the factors which prevented a concerted effort to encourage nationalization, as the demand for nationalization tended to be dropped during discussions over wages and conditions. It also meant that it was difficult to agree benefits and wages across the whole of the industry because the different heights of seams and methods of extraction affected productivity levels.

Probably the most public show of miners' disenchantment was the General Strike of 1926. The main reason for the strike was the massive deflation which occurred in the post-war period (Figure 3.1) which led to a decline in the profitability of the mining companies. They had been forced to accept lower prices for their coal as a result of this deflation, as shown by the more specific cost of living index for fuel and light and coal prices which are shown in Figure 3.2. This deflation meant that by the middle of 1925 some two-thirds of the coal mines in the UK were operating at a loss and a quarter of the workforce was unemployed. In order to take account of post-war price levels, cut operating costs and thereby restore profitability, the mining companies' asked the miners' unions to accept a pay cut. However, the miners were in a

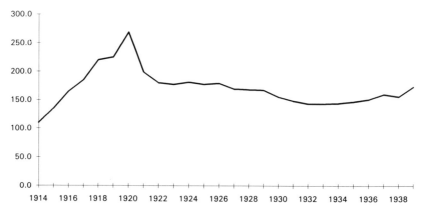

3.1 Cost of living index 1914–39
Source: HMSO.

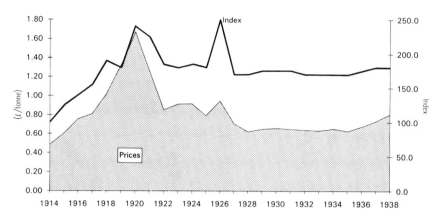

3.2 Coal prices and cost of living index for fuel and light 1914–38
Sources: Mitchell & Deane and HMSO.

very strong position at the time because they were still revelling in their success at winning the pay confrontation in 1920 and smarting from their failure to prevent the wage cuts in 1921. In 1920 the mines were still under government control and the country was faced with a potentially critical situation due both to low coal stocks and an absence of emergency powers which could have eased the situation. Initially the Lloyd George government had hoped to restrict the wage claim but, as a result of support from the railwaymen, which could have exacerbated the situation, it backed down and conceded a substantial rise. As a result an Emergency Powers Act was introduced to Parliament later in the year, which gave the government sufficient power to handle a national emergency.

Over the following years the situation between the miners and the coal owners returned to a relative calm, with the market for coal helped by problems at overseas producers. However, by the middle of 1925 disputes over pay and conditions were starting to rematerialize and a strike was again threatened. In this instance the government was able to avert a strike by providing a subsidy to maintain pay until the end of April 1926 and by setting up a Royal Commission to look into the miners' grievances. One of the reasons behind this dispute was the 1 May 1925 re-fixing of sterling to the Gold Standard by the Chancellor of the Exchequer, Winston Churchill, at the same rate (£3 17s 10d/oz) that it had been when sterling was released from gold at the outbreak of the war in 1914. This overvaluation of the pound meant that British goods were disastrously uncompetitive in the international market because of the high cost of buying the sterling in which they were priced.

At the time the export market for coal, whilst it had been adversely affected by the war, was still a major area of demand as it was British coal which was used by all of the British ships plying the seas. British coal was also shipped out to depots around the world for these ships to refuel and the opening up and expansion of coal mines in countries such as Australia meant that British mines had to reduce prices sharply in the attempt to maintain the supply contracts. J M Keynes calculated that the average profit on the sale of coal was some 6d/t during the first quarter of 1925 and that the costs of production had to be reduced by 1s 9d/t so that the same profit margin could be retained on maintained exports. This was only believed to be possible through a reduction in miners' wages. A J Cook, who was an extreme left-wing leader and general secretary of the Miners' Federation of Great Britain (MFGB), coined the slogan 'Not a penny of the pay, not a minute on the day'; but in the end it was to little avail.

The Royal Commission set up by the government to look into the miners' grievances was chaired by Sir Herbert Samuel. The four member Commission reported in March 1926 and agreed that wages should be reduced although it also pressed for a comprehensive reorganization of the industry. On 24 March the government announced its intention to give effect to the proposals of the Commission provided that both sides of the industry accepted them. The owners did not want reorganization imposed upon them any more than the miners wanted a pay cut. Much negotiation followed but both sides were intransigent and by 23 April the best offer which had been put on the table was a pay cut of 13% and an increase to an eight hour day for face workers. With the lack of agreement the miners were locked out from 1 May 1926. Three days later, at midnight on 3 May 1926, the General Strike began after the trade union leaders had met the Prime Minister Sir Stanley Baldwin and failed to reach an agreement.

Nine days after the start of the strike the Trades Union Congress called off its part in the action because it believed that this was then the only way to make any progress in the negotiations, and the miners carried on alone. It led to a great feeling of bitterness between the miners and their trade union comrades, especially as the dispute continued for seven months of protracted negotiations between the unions, the owners and the government. By early September 1926 only some 5% of the miners had returned to work, although by late November, in a mirror image of a later and now more well known event, the Nottinghamshire and Derbyshire miners created a breakaway union which made its own arrangements with the management and returned to work. A special miners' conference was held in mid-November at which time the leaders decided to recommend acceptance of the government offer and by the end of the month work had resumed at most coal mines.

The following 10 years gave the control of the union back to the local

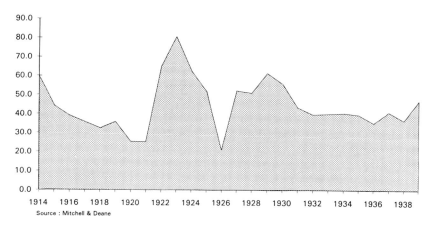

3.3 Coal exports 1914–39 (million tonnes)
Source: Mitchell & Deane.

level and most wage negotiations were carried out on a pit by pit basis. This often led to vast disparities in the amounts miners were paid for undertaking basically the same tasks in neighbouring mines. These disparities were sometimes reduced when miners transferred between pits, and it was in local disputes over these disparities that Joe Gormley, later a President of the National Union of Mineworkers (NUM), first started to press for better pay and conditions.

Unfortunately, the reduction in export prices facilitated by the wage cut was insufficient and overseas competition gained a greater share of the international market. As a result, the level of exports following the strike never regained the post-war peak of 1923 despite a strong rebound in 1927 as production levels were rebuilt. Thereafter the decline in exports shown in Figure 3.3 put further pressure on the industry to cut costs and reduce production to remain profitable. The depression, of course, also affected the industry and during the period miners suffered from the run-down of the workforce as a result of the decline in demand and the increase in mechanization. This meant that by 1928 some 33% of all registered miners were unemployed and by the depths of the depression in 1932 this had risen to a massive 40%.

The other reason for transferring power back to the local level was that the union, and indeed many miners themselves, had little desire for further confrontation – especially as the industry was contracting in response to the general economic malaise. As an example, Joe Gormley's first job in 1931 was in the Wood Pit in St Helens, but although the mine was owned by Richard

Evans Company it was under the control of the bank. The first 3d Gormley spent from his first pay packet was to join the union, and over the succeeding years payments into union funds helped to restore confidence. In 1935 the miners wanted an extra 2 shillings a day for their work. They balloted for a strike, which was approved, and gave notice that it would start on Monday 27 January 1936. In a peculiar show of foresight, which the electricity industry today is unlikely to repeat, several large coal consumers said that they would pay more for their coal supplies if the miners received the extra amount exclusively. Whilst the figure offered worked out at less than the 2 shillings a day that the miners were after, the pay increase was accepted and the strike averted.

World War II also saw periods of industrial unrest within the coal industry. Again it was a period of high inflation and, as part of the war effort, the miners were encouraged to work harder and for longer in order to maintain output. The miners wanted more money as compensation for the extra work that they were being asked to do and, having failed to obtain it, they went on strike in 1942. The strike started in Yorkshire and soon spread elsewhere in the country. In Kent 1017 miners at the Betteshanger Colliery were called before local magistrates for not giving the required wartime notice. The dispute was eventually settled and the miners returned to work.

PRODUCTIVITY AND MECHANIZATION

The other area of disenchantment within the industry concerned the appalling lack of productivity and the lack of investment in equipment which would boost the output of the mines. The slow move to mechanization is reflected in Figures 3.4 and 3.5, which show the development of mechanical cutting equipment and mechanical loading equipment. By the end of World War I the use of machines in British coal mines had fallen far behind their use in other countries. And even during the period until the end of World War II the introduction of coal cutting equipment in the UK had been much slower than elsewhere.

Obviously the problems faced by the mining industry during the depression of the late 1920s and the early 1930s were partially to blame for the lack of advance through the period. Nevertheless, the USA also suffered considerably during the downturn but was still able to increase its use of mechanization. The less advanced state of the mining industries in other countries must

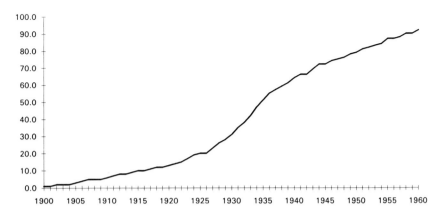

3.4 Percentage of coal cut by machine 1900–60
Source: Mitchell & Deane.

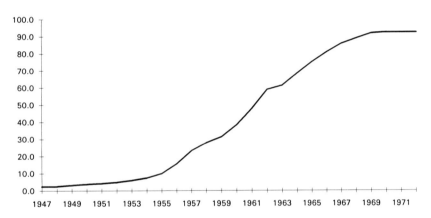

3.5 Percentage of coal cut by machine and power loaded 1947–72
Sources: Mitchell & Deane and British Coal.

be part of the reason for this as the Americans were able to design new mines employing all of the latest techniques, rather than having to wait for the depletion of old mines before these techniques could be implemented. It must also be the case that the British dependence on international trade and the contraction of trade during the depression (Figure 3.6) had a greater effect on British industry than on the more self-sufficient American economy.

The other problem was, and still is, an accounting one whereby all of the investment in plant and equipment in the UK is written off over the period of

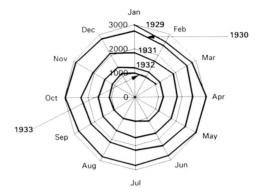

3.6 International trade contraction shown by total imports of 75 countries 1929–33 (millions of old US gold dollars)
Source: C Kindleberger, after League of Nations 1934.

useful life of the equipment. This means that there is a marked reluctance in industry to dispense with equipment which still works even if it is not as productive, and hence as profitable, as a new design of machine employed by a competitor. German depreciation policy, for example, is such that all new capital expenditure is written off in the first year of operation (together with the costs of retraining the workforce) and there is therefore little management concern about the carrying value of an asset in the balance sheet. Companies, if they have sufficient funds to afford it, can therefore order a new machine even before the first one is a year old without fear of the adverse effects the disposal of the old machine will have on the profits reported by the company.

The 1936 report prepared by the Political and Economic Planning Institute (now part of the Policy Studies Institute) on the UK coal companies compared the productivity of coal mines in different parts of the world between 1913 and 1934. In the UK productivity, shown by output per manshift, improved by 7% whereas in Poland it increased by 63% and in the Ruhr coalfield by a massive 77%. One of the reasons for the sharp increase in productivity in the Ruhr was the move over to mechanical extraction of the coal where some 97% of the output was won by machine in 1934 against only 22% in 1913.

The poor productivity performance in the UK is obviously related to the lack of mechanization in the industry. The use of pit ponies was an important productivity improvement of the middle of the eighteenth century, as a pony was able to pull much greater loads underground than a man. The average ratio of ponies to men was about 1:25 until after 1925 when the use of mechanical transport started to increase. There were over 70 000 ponies in use in the mines in 1912 and by 1924 this figure was still some 65 210. After then the numbers fell slowly, reaching only 32 524 in 1938 and 15 858 in 1951; by 1970 they had been eliminated from all but a handful of British mines.

The move to mechanical cutting of coal was restricted by the accounting policies and low levels of profitability outlined above. Nevertheless, the introduction of new technology did proceed, albeit slowly, and was broadly successful at increasing productivity. In 1938, when only about 56% of the total amount of coal cut in English and Welsh mines was cut by machine, about 68% of Scottish coal was machine-produced. The reason was that in Scotland face labour costs were relatively higher than they were in England and Wales and the Scottish mine owners therefore had greater incentive to mechanize their operations. The main reason for this is that Scottish seams are relatively thin and the mines had to use bord and pillar methods of extraction rather than the more productive longwall methods. The great advance in productivity provided by new technology did not occur until World War II when the AB Meco Cutter–Loader was developed. This machine reduced the need for men at the mining face from over 30 to between 13 and 15 and thereby led to a sharp improvement in productivity. The machines took some time to install because they needed to be placed in faces designed specifically to take them, and by 1948 there were only between 40 and 50 in use in British pits.

The use of underground lighting had not advanced far from the early nineteenth century invention of the safety lamp. Indeed, the Miners' Lamp Committee found in 1924 that there were 403 000 Marsaut-type, 17 600 Muesler-type, 6000 other flame and 328 000 electrically-powered safety lamps in use in 1923. Electricity had, in fact, first been used underground in the Earnock Colliery at Hamilton in Scotland and at the Pleasley Colliery in Derbyshire in 1882, but it took a long time to spread through to the rest of the industry. One of the main reasons for this was the danger of early electrical installations which could often generate sparks. These sparks could ignite any methane in the pit as easily as a naked match and electric lamps were therefore not as popular as safety lamps which, if operated properly, could not cause an explosion. Nevertheless, the Committee found that miners often failed to pay enough attention to their lamps and allowed the surrounding gauze to get too hot, thereby increasing the likelihood of an explosion.

Nevertheless, the increase in safety awareness in the twentieth century brought about many advances in the safety of miners. The average rate of deaths underground was about 216 per 100 000 employees in 1885 but this had fallen to only 62 per 100 000 employees in 1949. By 1970 the figure had improved still further to 30 and is now only about 1.5–2 per 100 000 manshifts. This is significantly better than most of the competitive mines elsewhere in the world.

One of the most ridiculous problems which adversely affected the British coal industry in the run-up to nationalization was a disagreement between the railway companies and the coal producers. This disagreement centred on the proportion of the cost of new, and more efficient, railway equipment which was to be borne by the two parties. By this time the railway network had spread across the country and it was easily the most efficient means of transporting bulk goods over long distances. However, whilst the network was complete it was owned and managed by different companies which operated to provide a transport service in their own area. As a result the transport of coal would have been difficult to organize unless the producers had their own wagons which could travel from one railway company's track to another's.

Engineering advances during the nineteenth century increased the power of steam engines and meant that they were able to haul larger trains. In order to take full advantage of this it would have been sensible to increase the size of wagons so as to cut down their unladen weight. This was widely accepted by 1914 but it was also understood that both the railway companies and the coal producers would benefit from the introduction of these larger wagons. In the case of the railway companies this was because a single engine would be able to haul more coal, although this would also help the coal producers because their output would reach the customer more quickly. Finally fewer, larger wagons would be easier to operate than many smaller ones.

The inability to obtain an agreement over who would bear what proportion of the cost of new wagons must have been exacerbated by the large number of companies on both sides of the negotiations. Indeed, it was not until the coal and the railway industries had been nationalized in 1947, and the government was negotiating with itself, that any agreement was reached. As a result the continuation of high transport costs during the inter-war years would have further reduced the international competitiveness of British coal. Moreover, the consequent high energy costs for British industry would have adversely affected its export potential, in much the same way that it is being damaged by high electricity prices today.

FINANCE AND CAPITAL

The build-up of capital was relatively slow and even by 1919 only about half of the 1452 coal mining companies which owned the 3300 collieries had converted to public companies. Of these companies there were 434 which produced less than 2000 t annually. Nevertheless, the average mine had some 300 employees in 1914 which, by contemporary standards, was quite a large concern.

One of the first companies to take advantage of the limited liability legislation was the Powell Duffryn Steam Coal Company Limited which was formed in 1864. Whilst the company went through a very poor period in the 1880s and was nearly wound up in 1888, output, helped by improving trade, grew over the years until by 1919 it employed over 18 000 men and produced in excess of 4 mt of coal a year. Many companies took the route of Powell Duffryn and expanded into related industries. Powell Duffryn was, and still is, interested in engineering which was originally a downstream part of its coal mining interests. At the time of nationalization the company had increased its operations to control some 65 mines employing almost 37 000 men and producing about 15 mt of coal a year.

If they did not move into steel and engineering the companies expanded into areas such as brick, lime and salt production as in the Skegby Colliery Brick and Lime Company Limited. This downstream activity also enabled the companies to find a market for the small coals they produced which would not find a ready market elsewhere in the industry. The final area for expansion, which had much to do with the companies' origins, was farming. The disposal of waste necessitated the ownership of surface rights as well as underground mining rights and the land was used for grazing the pit ponies. If the companies owned a large amount of land this could be used for farming on a larger scale.

The move into downstream activities did not find favour with the miners or their unions. This was because of the possibility of transfer pricing by which companies were able to take the profit from their mining operations in one of a number of other subsidiaries. As miners' wages were often linked to the profitability of the mining operation their disaffection arose due to the lack of transparency in the calculation of the profit of the mine in which they worked.

One of the particular areas of contention which exacerbated the miners' feelings of exploitation was the seeming inability of the coal price to decline during the period (see Figure 3.2). The main reason for this seems to have been that falling productivity in the mines meant that from the peak of production in 1913 the level of costs per tonne produced had increased, despite the reduction in employee numbers. In 1922 J W F Rowe mentioned, 'this predominance of demand explains the apparent paradox of rapidly increasing wages combined with a fall in the productivity of labour, and it has enabled the industry to pass on to the consumer the burden of that decreased productivity' (Griffin, p 126/7). When the fall in demand led to coal price cuts, and the deflation of the 1920s also started to eat into coal prices, the miners suffered a double blow which led to many of the unemployment problems outlined above.

GOVERNMENT INTERVENTION

The slow progress of government to control much of the coal industry of the UK gathered momentum during the first 50 years of the twentieth century. This was necessitated by the need to keep as much of industry as possible in operation and to attempt to reduce the debilitating effects of strikes and union militancy in the country. The first truly national strike in the coal industry was in 1912 and was sparked off by a pamphlet by a South Wales body, the Unofficial Reform Society (URS), which advocated an all-inclusive union and workers' control of the industry as well as calling for a national minimum wage. This demand was the main reason for the strike which lasted from the end of February until early April. Britain's dependence on coal as a source of energy at the time meant that the government was forced to intervene.

Asquith introduced the Miners' Minimum Wage Bill to Parliament in 1912 and is reported to have wept. The fixing of minimum wages was seen as the first step on the path to ruin by officials at the time. Nevertheless, the first true government intervention in the industry had occurred in 1893 in response to the national coal lockout. This intervention continued whenever the national interest was deemed to be at stake and led to the mines being placed under government control during both of the World Wars. The widespread nature of the electricity industry also led to the partial nationalization of that industry in 1926 and the government created an effective monopoly in the steel industry in 1932, in response to the problems of the depression.

A private cartel was put together by the coal producers at the end of the 1920s. This cartel imposed a quota and price system through a central selling organization in an attempt to control the market. The operation of the cartel was eased by the Coal Mines Act of 1930 and it was helped further, and given government backing, in 1936 following the threatened strike of 1935. The agreement of the main coal consumers to accept higher prices if their supplies could be guaranteed was part of the arrangement. Prices and output of coal were then controlled by government and led to the proposal for government ownership of the coal royalties in the 1938 Coal Act which was eventually implemented in 1942; nationalization in all but name.

However, despite the government's wishes there was no concerted effort to restructure the industry until the Labour government gained power following the end of the war in 1945. There were many calls during the period for a total reorganization of the industry emphasizing the need for the amalgamation of companies in adjacent pits in order to attempt to reduce overhead costs and improve productivity by working pits as single rather than dual operations. All the studies and investigations continued on the basis of the recommendations originally put forward in the 1919 Sankey Commission report which suggested that nationalization was the only way to force the restructuring of the industry. The Samuel Commission, which agreed with the coal companies' demands for a wage cut in order to return the companies to profitability, also stated that inefficient pits should be eliminated. At the time only 69 of the 1480 companies which were then in existence produced more than 1 mt of coal annually. This compares with much higher numbers in both the USA and Germany which were Britain's major coal competitors at the time.

THE ENERGY CRUNCH

Coal also suffered during the period from the end of World War I because of the massive increase in oil and natural gas production which was being used to meet the world's growing energy requirements. This can be shown by the increase in world coal demand which averaged some 4% a year before 1913, but which fell virtually to nothing in the post-war period until the start of the 1930s.

Following the end of the war the control of the Middle East fell largely to the British. This included the vast oil wealth of the area and the early growth of both British Petroleum (BP) and Shell can be traced to their interests in the region. The greater influx and relative cheapness of oil led to a realization that Britain did not need to be as dependent on coal as it had been in the past. The general rise in mobility also increased the demand for oil products for transport as more motor cars became available. The import of oil also brought about a change in the shipping capacity during the inter-war years with a much greater emphasis being placed on oil-fired rather than coal-fired ships. This affected the demand from the shipping industry for coal for use at home and for export, and also led to a decline in the demand for coal for use in the steel manufacturing industries as the overall demand for shipping capacity declined.

This, admittedly, was not just a feature of the UK energy balance but was also the case in other countries around the world and was a further factor in reducing the export market for the British coal industry. Total world coal production increased from around 135 mt in 1860 to 700 mt in 1900 and some

4539 mt in 1990. This, though, must be put against an increasing demand for other forms of energy around the world which resulted in coal's share of world energy consumption falling from an effective 95% in 1900 to only 27% in 1990.

This declining market share has been caused by a realization of the relative costs of the respective energy sources. It is clear that the cost of digging mines underground and employing many men must be greater than employing a few men to drill and pump a massive oil- or gas-field. The downstream refining costs of producing petrol or other fuels are necessarily greater than for coal because coal simply needs to be washed to remove dirt in order to attempt to reduce its ash content. However, if coal was to be used to manufacture hydrocarbon fuels the costs would be greater than those associated with cracking crude oil. This is because of the very low concentration of hydrogen in coal. When one day the price of oil has been forced to such heights by its scarcity it may be the case that relatively cheap hydrocarbons can be produced from coal. And, as Joe Gormley said to David Howell, then Minister of Energy (in February 1981), 'once it becomes commercially possible to produce gas and oil from coal, no one is going to want to export coal'.

The move away from combined to individual methods of transport and the advent of the motor car, in particular, have meant that the worldwide spread of energy consumption has also changed, with proportionally more being used for transport than for basic heat and light (both through gas and electricity generation) than in the past. Nevertheless, the dependence on coal has been declining in response to cheap alternative sources of energy and although the inter-war years saw the start of this decline it became most apparent during the 1960s.

Chapter
4

Buying the
family silver

MOTIVES AND METHODS

The increasing acceptance of Keynesian economic theory during the 1920s and 1930s occurred at a time when governments, of whatever political persuasion, were becoming increasingly concerned about the poor state of the mining industry. A general belief that the British mining industry would never become internationally competitive, unless it was forced to change, gained credence, especially following the Samuel Commission report in 1926. It was further enhanced because coal was then still viewed as an important national asset, which was needed in times of war and therefore had an important role to play in ensuring the security of the nation.

The arguments that the coal mining industry should be nationalized were first put forward by the mining unions in the nineteenth century. This was in furtherance of their demands for better pay and conditions and they were behind the commitment for nationalization adopted by the Trades Union Congress in the 1890s. This was especially enhanced by the high mortality rate with over 1000 miners killed each year in the thirty years between 1856 and 1886. Such a proposal was dismissed out of hand by many of the mine owners but, in 1893, in response to the national coal lock-out, Sir

George Elliott advocated the formation of a coal trust which would attempt to rationalize the industry. Many of the smaller and inefficient mines believed that they would be undervalued in such a compulsory reorganization and they campaigned successfully against its implementation. Nevertheless, the first nationalization bill to include a proposal to nationalize the coal mines was the Nationalization of Mines, Canals, Railways and Tramways Bill No 103 which was put before the House of Commons in 1906. The same bill was reintroduced in 1908 but omitted the clauses referring to nationalization of the mining industry. A further Nationalization of Mines and Minerals Bill was proposed in 1913. None of these bills made any real progress in the House.

Despite these set backs the arguments for nationalization slowly began to percolate through many sections of government and society. They were helped by the control of the mining industry which was exercised by the government during World War I and the problems encountered following the return of the mines to private control in 1921. These problems continued for a long period after World War I and were exacerbated by the depression of the early 1930s which forced the government back into the fray in order to attempt to control a national resource. The Labour Party, was, however, always in favour of the nationalization of the industry and endorsed a policy document, 'Labour and the Nation', at its 1928 conference which promised the nationalization of coal among other industries. Even though Britain was slowly moving away from its dependence on coal as a source of energy, its widespread use meant that the economy would still grind slowly to a halt if it was not oiled with coal.

During the late 1920s and early 1930s the level of government intervention slowly increased and the private companies resented this as much as the government saw the need for it. A number of Royal Commissions had been set up to look at the coal industry in detail in order to attempt to determine how it should continue and how the future of the country's energy supplies could be guaranteed. The main conclusion of these studies was that there needed to be a concerted shake-up of the industry with the emphasis being placed on the creation of fewer, larger and more productive operations which would benefit from economies of scale. In the 1930 Coal Mines Act a Coal Mines Reorganization Commission was established in order to promote the reorganization of the industry through the amalgamation of individual collieries into larger operating units. However, strong opposition from the industry prevented the Commission from having much effect on the structure and control of the mines.

One of the most obvious halts to progress, however, must have been the level of government involvement in the industry during the war years. This was because the amount of investment in new plant and machinery must have been reduced because of the need of the manufacturing sector to produce armaments in place of coal cutting equipment. Additionally, the mine owners would probably not have wanted to invest in new equipment which would be unlikely to generate a direct return if the mines remained in government hands for any length of time. This lack of investment was apparent from the conclusions of the various Commissions and, in much the same way that the 1984/5 strike delayed the move to greater productivity, must have had damaging effects on the longer term output of the industry.

As a result it slowly became clear to the government that the only method to guarantee the necessary amount of investment in new equipment would be if the coal industry was under state control. In this way the government would be able to plan for the future and ensure that the country would have sufficient supplies of energy to last, in political terms at least, forever. A great debate about the future of coal was conducted in the run-up to World War II and the final outcome was the 1938 Coal Mines Act. This Act, which became effective in 1942, provided for the vesting of all coal in the country into a new body, the Coal Commission, which would regulate the industry and earn royalties based on coal production.

LABOUR SWEEPS HOME

Labour was swept to power in the general election following the end of World War II. The government, led by Clement Atlee, had been elected on a wide-ranging programme of nationalization and the wheels were set in motion for this soon after the results of the election were announced. In particular the Labour Party election manifesto had stated that

> *for a quarter of a century the coal industry, producing Britain's most precious national raw material, has been floundering chaotically under the ownership of many hundreds of private companies. Amalgamation under public ownership will bring great economies in operation and make it possible to modernize production methods and to raise safety standards in every colliery in the country.*

Table 4.1 shows the distribution of these companies and their mines by county.

One of the problems the manifesto did not address was who was going to pay for the massive nationalization programme which had been set out in it. For it was not just the coal industry which was to be nationalized; the Bank of England, the railways, and the electricity and gas industries were all scheduled to come under government control and this meant that vast amounts of money had to be raised. The broad scope of the programme was partly a matter of policy, but also reflected the Labour Party's fear that the 1950 election would see a return to a Conservative administration which would then

Table 4.1 The distribution of mining companies and their
mines at the end of World War II

County	Companies	Mines
Northumberland	38	79
Durham	55	152
Cumberland	11	16
Yorkshire	107	184
Derbyshire	39	83
Leicestershire	10	11
Nottinghamshire	18	40
Warwickshire	15	16
Lancashire	42	94
Denbighshire & Flintshire	11	12
North Staffordshire	18	23
South Staffordshire & Worcestershire	33	47
Cheshire	3	3
Shropshire	10	14
Somerset	7	11
Gloucestershire	11	14
Kent	3	4
Monmouthshire	27	55
Glamorganshire	46	142
Breconshire	4	9
Carmarthenshire	7	32
Pembrokeshire	1	1
Clackmannanshire	2	11
Fife	12	35
Haddingtonshire	3	7
Lanarkshire	32	79
Dumfriesshire	1	4
Midlothian	11	23
Dumbartonshire	3	7
West Lothian	13	26
Renfrewshire	13	19
Stirlingshire	4	7
Ayrshire	11	38
Sutherlandshire	1	1

Source: Anderson, Coal: A Pictorial History.

seek to reverse many of the moves Labour had made. In fact Labour won the February 1950 election but with a reduced majority of only 6 seats. They were then forced to call a second general election for August 1951 which gave the Conservatives, under Sir Winston Churchill, a clear majority of 17 seats.

This lack of funding led the government to attempt to place as much of the burden of nationalization as possible on to the industries themselves. Some were unable to cope but the coal industry was forced to take on debt through the payment of cash compensation. This method of funding the acquisition was used in order to attempt to reduce the amount of new gilts issued so that the government could continue to fund its other, more socially oriented, pro-grammes. Unfortunately, it meant that the new National Coal Board did not have access to sufficient capital to reorganize the industry as quickly as it may have liked. The post-war growth in government spending also germinated the seeds sown when sterling was taken off the Gold Standard on 21 September 1931 and led to the massive inflationary boom of the post-war years – a boom from which the country has yet to recover.

The other problem faced by the coal industry was the shortage of labour. This may seem strange because of the return of soldiers from the front line looking for employment after the end of the war. However, the mines did not lose many workers to the armed forces during the war because their industry was deemed to be a national priority and the production of coal had to be maintained in order to help the war effort. As such the mines only lost some 50 000 employees in World War II compared with the 250 000 who joined up in the first year of World War I. After the end of the war the use of outside labour diminished as people who had come into the industry from outside in

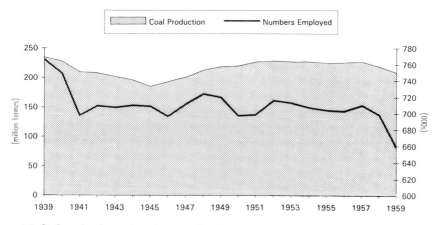

4.1 Coal production and numbers employed 1939–59
Source: Mitchell & Deane.

order to help in the war effort left to rejoin their original careers or returned to their former countries. As a result the numbers employed in the industry actually fell in 1946 although this did not appear to have an adverse effect on production (Figure 4.1). The need for more labour was realized and a recruitment campaign was initiated with the result that a net 27 000 people had joined the industry by 1948. Numbers, however, soon started to tail off and although this was helped to a certain extent by increasing productivity the peak post-war output of 230 mt in 1952 was still lower than the 1939 figure of 235 mt.

ACTS AND ACTIONS

The nationalization of the coal royalties has been covered above but it did not meet the expectations of its proponents and the industry remained inefficient and uncompetitive. A 1945 report by Sir Charles Reid stated broad agreement with the structure of the Coal Commission and confirmed that it, or another body planned along similar lines, should ensure that the industry was merged into organizations of such a size that production would be improved. Whilst it obviously called for state intervention in the industry it did not specifically advocate its nationalization.

One of the beliefs which started to permeate the Labour party at the time was that the miners would only work harder and for longer if they knew that they were working for the state rather than for the benefit of individual companies. However, the efforts of the miners to increase output during World War II often led to a reduction in safety standards and the reintroduction of such standards and the knowledge that there was not a war to be won meant that the miners were less inclined to increase production at the expense of longer and more dangerous working. As a result such socialist ideals did not lead to increased output despite the continuation of left-wing leadership of the union. The miners themselves were just interested in more pay and shorter, safer working hours.

Out of necessity the first nationalization of the Labour government was of the Bank of England. This was because it needed control of the Bank in order to ensure its ability to fund the acquisition strategy outlined in the manifesto. For the coal industry the nationalization bill was first published on 20 December 1945, it had its second reading on 29 January 1946 and was given Royal Assent on 12 July 1946. This gave the effective date of the nationalization as 1 January 1947.

As part of the nationalization process the newly established National Coal

Board (NCB) took over all of the assets of the 1938 Coal Commission. This gave the NCB control of all of the coal deposits in the country and effectively restricted the formation of any private operations which could match the NCB in size and scale of operation. One of the reasons was that these operators were forced to pay a royalty to the Coal Commission, and subsequently to the NCB, and this increased the effective cost of production, reducing profit margins. Because it was not efficient for the government to nationalize the entire industry down to the smallest mines, those private companies which employed fewer than 30 underground workers were excluded from the nationalization process.

In 1945 there were 521 mines employing fewer than 30 people. This represented around a third of the total number of mines in the country but accounted for only about 1% of total coal output because of the small scale of the operations. By the time of nationalization in 1947 there were 480 private mines which were excluded from being nationalized on this basis. Further, they were prevented from increasing their output following nationalization by the introduction of a maximum limit on the number of underground workers they could employ. In order to remain in line with the nationalization exclusion this limit was kept at 30.

Initially there was much discussion about what methods should be adopted in order to ensure that the owners of the 980 larger mines were paid a reasonable price for their assets. There were four specific possibilities which were as follows:

1. On the basis of reasonable net maintainable revenue.
2. On the basis of Stock Exchange quotations.
3. On the basis of total capital expenditure less accumulated depreciation.
4. On the basis of a valuation of the companies' physical assets.

Of these the first option was chosen because there were seen to be too many problems with regard to the other methods of valuation. First, not all of the coal mining companies were quoted and therefore a valuation of an unquoted company could not be obtained. For the larger, more diversified companies it was also difficult to determine how much of a company's market capitalization was related to its coal producing assets and how much to its other assets. This was further complicated because these other assets could be in any sector of the economy. As is often the case with balance sheets, it was argued that the total capital expenditure less the accumulated depreciation of a mining company was not indicative of its true worth. This could arise if the company had acquired many of its assets some time before the balance sheet was drawn up. In this instance a company could have depreciated them over a considerable period of time, thereby reducing the amount shown in the balance sheet. Such a figure would be unlikely to reflect even the value of the coal reserves, let alone the additional infrastructure and assets. Finally, it was considered that a formal valuation of the assets of each company would take too long and would open the way for disagreement as companies would argue that the valuation was inaccurate as it did not take various factors into account. The first option was, therefore, seen to have fewer problems than the others especially because the nationalization of the coal royalties in 1942 had been on a similar basis and had, therefore, set a precedent.

The nationalization of the coal royalties cost a total of £76.45 million (£1476 million in 1993 money). This figure was arrived at as £10 million plus the value of the royalties as then calculated using current production statistics as a method of forecasting the future production on which to base the royalties. As sterling was still holding its value against all other currencies, despite the devaluation of 1931 (almost all other major currencies had been devalued at a later date), no account was taken of the cost of capital or of the time value of money through the use of discounting formulae to arrive at a present value of the royalties. Therefore the £66.45 million of royalty recompense was calculated on an annual average royalty of £4.43 million and the average life of a royalty of 15 years. The only problem this method provided for the full nationalization of the industry via the reasonable net maintainable revenue method was that the level of output and prices had been affected by the war. This led to some problems over the calculation of the amounts to be paid.

As the process was pursued, more problems arose and the original idea to pay out some of the compensation by way of an issue of fixed interest coal stock was turned down by the Minister of Fuel and Power. This was because he would have regarded it 'as disastrous if the financial arrangements were such as to render some faint colour to a suggestion that private individuals still owned the industry through their holding of coal stock, and that the industry had not been truly nationalised' (Chester, *The Nationalization of British Industry*, p 241). As a result the government decided that the nationalization had to be funded by an issue of government stock. However, in order to attempt to reduce the effect of the issue on the gilt market the owners were paid in stock, rather than in cash, and a restriction was placed on the sale of the gilts in the secondary market. This is because it was considered that a major sale of gilts could have had a disastrous effect on interest rates and the performance of the currency on the foreign exchange markets.

The disagreements on the amount of money the government should pay in compensation then caused many headaches. The industry decided that the government was not offering enough to purchase the coal mines and was determined to obtain more money. The government reluctantly set up a

Table 4.2 The cost of nationalizing the coal industry in 1947

Items	£000 (1947 money)	£000 (1993 money)
Coal industry (terms of settlement) assets	164 660	3 082 310
Subsidiary (ancillary) assets	90 562	1 695 250
Minerals	80 888	1 514 160
Severance	689	12 900
Capital outlay refunds	16 774	314 000
Stocks of products and stores	34 631	648 260
Additional compensation	3 824	71 580
Total cost	392 028	7 338 460

Table 4.3 Costs incurred by the mining companies in the run-up to nationalization

Items	£000 (1947 money)	£000 (1993 money)
Stocks of products and stores	34 631	648 260
Capital outlay refunds	16 774	314 000
Railway wagons	12 300	230 250
Other	2 295	42 960
Total	66 000	1 235 470

tribunal to look into the amount of money which should be paid out and it arrived at a so-called 'global sum' for the nationalization of the industry. This sum amounted to £164.66 million (£3082 million in 1993 money) and was a total figure which then had to be split between the colliery owners in proportion to the amount of money they thought their individual operations were worth. This, of course, set the scene for further disagreements as companies attempted to prove that they should receive a larger proportion of the total.

The total payments made for the acquisition of the coal industry also included various extra amounts which, among other items, took capital expenditure in the year running up to nationalization into account. It further provided some recompense for the takeover of subsidiary assets and mineral stocks which were left unsold at the time of the takeover. These additional costs amounted to £227 368 000 (i.e. more than the specific cost of the acquisition). The amounts paid at the time and their equivalent in 1993 money are shown in Table 4.2. Further it should be noted that the National Coal Board also had to assume the government's debt incurred in connection with the acquisitions of the royalties in 1938. This figure was then calculated at £78 457 008 in comparison with the original cost to the government of £76.45 million.

Of the amount which was paid out a total of £66 million was in cash. This figure largely repaid the cost of the capital outlays and various other costs which the companies had incurred in the run-up to nationalization. These costs are detailed in Table 4.3.

A SECURE FUTURE?

The new National Coal Board could hardly have got off to a more inauspicious start. A member of the newly formed National Union of Mineworkers had stated at the union's annual conference in 1945 that 'without coal the economy will fall, and if the economy falls so will Labour'. However, in the early days of 1947, just after the nationalization of the industry had been completed, the economy very nearly did fall.

In the immediate post-war period there was a massive increase in energy demand from the country as it started to rebuild. The coal industry was struggling to maintain supply and had been lucky that the winter of 1945/6 had been relatively mild. The following year, however, was a different matter and the extremely cold weather forced up energy consumption as the population attempted to keep warm. This led to a severe shortage of coal which came to a crisis point in February. On Friday 7 February 1947 the Minister of

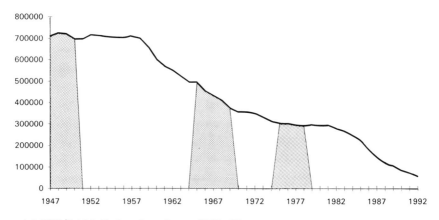

4.2 NCB/British Coal total employees 1947–92
Note: shaded areas represent periods of Labour government.

Fuel and Power took the unusual step of making a statement to the House of Commons late in the afternoon. This would have been when most MPs were already on their way home to their constituencies for the weekend. The statement announced that there would be no electricity for industrial users in London and Manchester from the following Monday, 10 February. If this measure had been unsuccessful the government had decided that the evacuation of London would be a possibility.

Whilst the year leading up to the nationalization of the industry must have led to a lack of investment in plant and equipment, together with a reduction in the number of new mining faces opened, these were not the only problems the newly consolidated industry had to face. To maintain, and even increase, production the industry also needed to retain its skilled workforce. However, the higher pay offered by other employers for much easier and less hazardous jobs led to a reduction in the numbers employed in the industry, a trend that has continued to the present day (Figure 4.2).

The long decline (1945–1979)

INVESTMENT AND RETURNS

After nationalization the decline of the British coal mining industry started in earnest. The blame must be laid partly at the feet of the Labour and Conservative governments of the immediate post-war years. Both parties failed to invest in the industry on the scale required both to boost productivity and to retain the experienced workforce. The other problem was the expansion in the availability and use of other fuels which led to a relative decline in the position of coal. Nevertheless, it is unclear whether the private coal owners would have fared any better as, undoubtedly, the main problem was investment. Coal needed lots of it but the Labour government, committed as it was to a massive nationalization programme, could not afford to provide the coal industry with as much money as it needed.

If coal had remained in private hands the attrition of the post-war years may have taken place at a faster rate than the rate of decline under government control. This would have been because the high level of losses across most of the NCB's divisions (see Table 5.1) could not have been sustained in

Table 5.1 The NCB's losses in the post-war years

Year	Profit/(loss) before interest (£m)	Interest (£m)	Surplus/(deficit) after interest (£m)
1947	(7.9)	15.0	(22.9)
1948	17.6	15.1	2.5
1949	31.2	13.4	17.8
1950	26.2	14.7	11.5
1951	18.5	14.7	3.8
1952	7.5	14.9	(7.4)
1953	22.7	17.1	5.6
1954	16.2	18.1	(1.9)
1955	2.0	21.3	(19.3)
1956	39.7	21.9	17.8
1957	19.9	25.2	(5.3)
1958	19.1	32.6	(13.5)
1959	13.1	37.1	(24.0)
1960	20.2	41.5	(21.3)
1961	28.6	42.4	(13.8)
1962	45.4	44.0	1.4
1963/4*	72.5	52.9	19.6
1964/5	42.8	42.7	0.1
1965/6	0.2	25.0	(24.8)
1966/7	28.5	28.2	0.3
1967/8	34.6	34.2	0.4
1968/9	28.6	37.5	(8.9)
1969/70	8.8	34.9	(26.1)
1970/1	34.1	33.6	0.5
1971/2	(120.2)	36.8	(157.0)

* 15 months.
Source: British Coal.

the private sector. Admittedly, these losses were calculated after charging interest on the NCB's borrowings from the government, a situation that would not have arisen in the private sector. However, the shortage of investment and equity capital in the 1940s and 1950s, and the lower investment that would have resulted, is likely to have led to a faster run-down of the industry, especially if the coal producers had attempted to push up or maintain the price of coal in the face of cheaper foreign imports. This had been the cause of the massive problems of the early 1920s.

The new investment that there was did go into many of the areas which had been neglected over the inter-war years and this led to a great improvement in the safety of the industry as a whole. The move to use more machines also helped to improve productivity but the improved productivity did not lead to a corresponding improvement in the pay and conditions of the workforce. The slow loss of labour turned into a flood and the industry found that it was unable to meet the production targets which were required for it to maintain its share of the energy market.

Following the end of World War II there was also a marked increase in competition from other sources of energy with increasing emphasis being placed on oil and, as the technology was developed, nuclear power. As an example, total world demand for oil ran at 5.2 million barrels a day in 1938, but by 1979 it was about 12 times this figure. This diversification of energy dependency came at a time when the coal industry was on its knees and in no state to compete. Therefore, the newly nationalized industry suffered because of the promotion of other forms of energy production. This promotion was often encouraged because the government could see that it was cheaper than increasing the amount of money given to the coal industry every year. And despite the massive injections of capital, the industry failed to make any money on a cumulative basis in the years up to the first national miners' strike in 1971/2. Obviously if interest payments are excluded the situation is very different but the cost of capital needs to be taken into account in order to provide a sensible comparison with the private sector. Whilst the left wing of the Labour Party may not have wanted to generate profits from the industry, without money being earned that same government would not receive any income return, nor would the industry itself be able to invest in new equipment unless its borrowings increased.

This increase in borrowings accrued as an obligation of the National Coal Board to the government and meant that the NCB had to pay the government increasing amounts of interest which reached a peak in 1962. The state of the market was also such that the company was unable to charge a price for its production which would generate a positive return and so it slipped into losses year after year. The government was eventually forced to write off some of the money but this, too, was insufficient to turn around the decline and return the mines to profit. This overhang of debt would not have been incurred so rapidly if the industry had remained in private ownership and it is unlikely that the companies would have fallen into the debt trap which the public company suffered. Nevertheless, it was clear before nationalization that the companies needed to invest a very large amount of money to improve productivity and safety records in the mines. They had been failing to do this and the move to nationalization was seen by many as the only method which would ensure that the British coal mining industry had a future.

MINERS AND THE MASSES

The militancy of the mining unions went through vast cycles of increase and decline between 1947 and 1979. The main reason for this was simply that the mining unions did not want to upset the apple-cart of a Labour government but did not care too much about the continuation of a Conservative administration. The old disagreements between the owners and the workers sprang up again despite the fact that the workers themselves were part owners of the industry and that it was not a Conservative pawn on the battlefield of British politics. The extreme militancy of the miners can be seen in the fact that in the first ten years after nationalization the coal industry accounted for one third of all days lost because of strikes and yet it only accounted for 4% of the British workforce.

There has been much speculation about whether or not the miners' strike of 1974 brought down the Conservative government of Ted Heath or whether it was the government's own policies which led to the fall. In fact the downfall of the Conservative government has probably less to do with the miners than it has to do with public disquiet over the parlous state of the country at the time (see below). This state had been engendered by the nationalization policies of the successive post-war Labour governments which put the country into debt which could only be financed by the massive devaluation of sterling which has taken place in the post-war period. The only comparable period in economic history was during the reign of Henry VIII when the currency was continually debased in order to finance the king's massive expenditure.

As such Britain in the post-war years has suffered from a declining currency and a continuing rise in inflation which has meant that workers have required more money in order to maintain their standard of living. The annual increase in the cost of living, shown by the rise in the Retail Price Index, is demonstrated in Figure 5.1. When compared with Figure 5.2 it shows that there was an increase in union militancy during periods of high inflation as miners attempted to maintain their income in real terms. Admittedly, this was true of all employees in whatever sector of the economy they were employed. The sad fact is that for many years the miners were given a relatively low level of wage increases and their standing against other workers in the economy slipped from the top of the pay scale to the middle. For the miners this was an untenable situation in view of the much greater amount of manual and often highly dangerous labour which they were asked to undertake as a normal part of their daily work. It was also a major factor in the reduction of the workforce during the period.

The formation of the National Union of Mineworkers (NUM) on 1 January

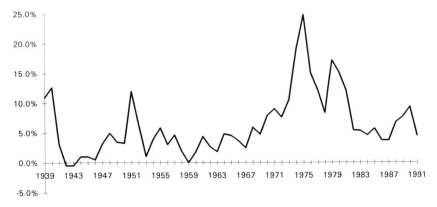

5.1 Annual change in RPI 1939–91

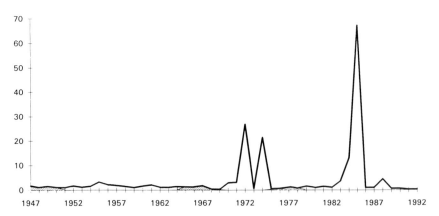

5.2 Tonnage lost due to industrial disputes 1947–92 (million tonnes)
Note: shaded areas represent periods of Labour government.

1945 also helped to consolidate the control of the miners into a single body with which to attack the government. The NUM effectively took over from the old Miners' Federation of Great Britain (MFGB) which had been formed in the late nineteenth century, and with nationalization the new NUM recombined the different branches of union membership which had come about following the split into separate areas after the collapse of the General Strike.

PIT CLOSURES

This is probably the most emotive issue of any discourse on the mining industry of this country. The problem lies with matching, or more accurately attempting to match, the supply of coal with the demand for it at the price at which it is produced. For the entire period following the end of World War II until the first oil shock in the run-up to the Yom Kippur War in 1973, coal was gradually becoming more and more expensive in relation to oil. This meant that the relative demand for oil was increasing at the expense of coal. On top of this the rise in clean air legislation and the fear that the polluting effects of coal burning would destroy the quality of life in towns and cities reduced the demand for coal for domestic heating, whilst the British export market was severely damaged by the war, as it had been during World War I.

As a result the governments of the time were forced to do the unthinkable – to attempt to reduce production. Indeed, it is strange to think that the reduction in the production of coal during the 1950s and 1960s was much greater than the fall in production which occurred during the Thatcher years, although the relative decline is similar. It is especially strange to consider that this reduction took place during continued periods of Labour government which, it might normally have been assumed, would have attempted to maintain jobs for miners in preference to producing profits for the Arab sheiks who were pumping so much oil.

This is also intriguing if the positions of both Harold Wilson as Prime Minister and Tony Benn as Minister of Energy are considered. These two MPs were very much allied with the fight of the coal workers and Harold Wilson, in particular, could trace his interest in Labour politics back to World War II when he worked as a statistician in the Board of Trade, and later the Ministry of Fuel and Power, obtaining information on the amount of coal which was produced during the war. One of the problems which he outlined was that the coal owners were making much of the statistics which showed that there was a marked increase in the number of shifts which were missed during the war. This, however, did not take account of the increase in the total number of shifts which were worked which showed that the relative number of shifts missed was not markedly different. However, it should also be considered that the miners were suffering from a lack of incentive to work harder during the war. As Wilson later recorded in his diary Will Lawther, the General Secretary of the National Union of Mineworkers, told him that

> absenteeism occurred in World War One due to the absence of goods worth working for in local shops. . . . After the war, when there were goods in the shops again, it would have taken a broken leg for him or one of his brothers to have a day off.

Wilson's Plan for Coal, published in 1945, outlined his vision for the coal industry and his belief that nationalization was the only way for the industry to succeed. This sits far from the decimation of the industry under his 1964–70 tenure as Prime Minister during which time more miners left the industry on an annual basis than during the five years of the previous Conservative administration. The total number of colliery closures in the 1967–8 financial year was 59 – an alltime record. Much, though, rests on the problems of the industry which arose out of the many years of private ownership in the inter-war years, and the reduction in the amount of capital investment which took place as a result of the uncertainty with which the industry was faced. It was also the case that many of the mines had been opened many years before and were therefore moving towards the end of their useful economic lives. This meant that mines had to be closed because there was no coal left in them. What was not done, however, was to replace the closing mines with new ones in order to maintain total output, and overall production started to fall faster than it had in the immediate post-war years when miners were tempted to higher paid, and safer jobs elsewhere in industry. In addition, the government started to reduce its dependence upon a single source of power as can be seen from the decision to build a nuclear power station in the middle of the Durham coalfield. The decision was made by Roy Mason, himself an ex-miner, and was received with disbelief by the industry.

The number of pits, the total output of coal and the number of miners working in the industry from 1947 to 1979 are shown in Figures 5.3, 5.4 and 5.5. These show how badly coal was affected by the problems the industry faced during the 1960s. These problems were further caused by the Wilson

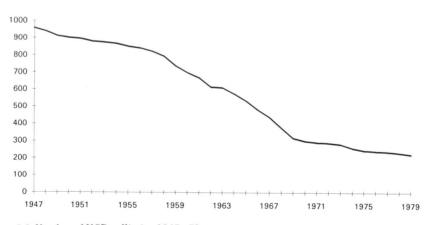

5.3 Number of NCB collieries 1947–79

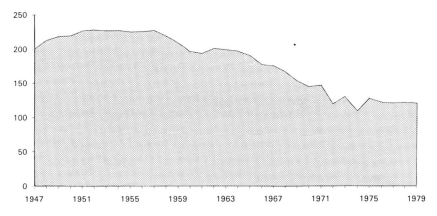

5.4 Total British coal production 1947–79 (million tonnes)

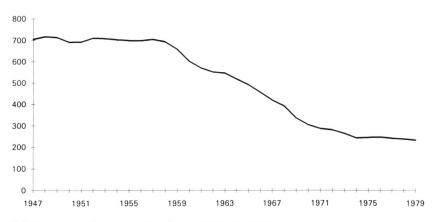

5.5 Number of miners on colliery books 1947–79 (000s)

government's financial problems during the period, culminating in the 1967 devaluation, and its unwillingness to write off its loans to the NCB. This meant that the Coal Board was essentially crucified by the interest charge which it was forced to pay to the government on an annual basis and this left it with even less money to improve productivity and output. The thought of an investment in a brand new mine, with the massive amounts of money which that required, would have been a fantasy dream for Alf Robens, later Lord Robens, who was Chairman of the NCB at the time.

1974 AND ALL THAT

One of the great sagas of mining folklore covers the time from the election of the Conservative government under Edward Heath in June 1970 until its demise in February 1974. The miners' struggle for more pay and better conditions was one of the continuing problems which faced the government, and was a legacy of the rampant inflation of the period (see Figure 5.1). This meant that the miners wanted continuing increases in pay in order to maintain their standard of living (much of which had been eroded because of their support for the previous Labour administration and their desire not to be seen to be rocking the boat by asking for large wage increases). However, the government wanted to put the message across to industry that large wage rises were inflationary and were therefore not in the best interests of the country.

The miners' leader at the time was Joe Gormley and he had many formal and informal meetings with the government in the attempt to get better pay and conditions for his members. The main problem was that the government refused to admit that the miners were a special case and that they deserved to have their relative standing in pay maintained. The first national strike the union called was in 1972 and its aim was to restore the pay differentials of the miners which had fallen to the extent that miners were in the middle of the wage rate table, in spite of the great danger of their occupation, rather than being at the top. This was the first national strike in the history of the NUM, although there had been many disputes before and the previous national union, the Miners' Federation of Great Britain, was, of course, involved in the General Strike of 1926.

This strike started at midnight on 8 January 1972 and ended on 28 February 1972 with the government capitulating to the union's demands. The strike saw the start of more concerted action by the unions in order to attempt to make their strike action more effective. This initially centred around the use of picketing at many different sites around the country in order to attempt to stop the movement of coal. This was necessary because the government had been building up stocks of coal at different power stations and depots rather than leaving it in stockpiles at the pits. A constraint on the transport of the coal was needed in order to reduce the supply of coal to power stations and in order to make the strike more effective. The employers also started to use non-union employees in order to attempt to break the strike, as these people would have no qualms about breaking through picket lines. In an attempt to counter this the unions then started to utilize mass picketing as a technique to prevent the movement of coal, and the organization of mass picketing was one of the areas where Arthur Scargill first came to prominence. The day that the strike ended also saw the introduction of the anti-strike provisions of the Industrial Relations Act which attempted to curb union activity.

This pay level was satisfactory to the miners for the next two years but by the time the negotiations for the 1973/4 wage rise started in late 1973 the miners calculated that their relative earnings had again declined. Joe Gormley had many secret meetings with Edward Heath and thought that he had provided the government with a get-out clause for the miners to be excluded from Phase Three of the Incomes Policy. This Policy was to be introduced in order to attempt to reduce the rampant inflation of the period (which was seen to be caused by a wage-pull rather than a supply constraint problem in the economy) through legislation in much the same way that the Major government elected in 1992 is attempting to control inflation through the encouragement of wage restraint in the private sector. However, when the Incomes Policy was presented to Parliament it was seen that the clause, which covered the working of unsocial hours, could apply to all workers, particularly those in the state sector who worked in shift employment. As its scope could not be restricted it would not have enabled the miners to regain their pay differentials and disquiet in the industry began to resurface.

The NCB was further suffering at the time because there were some 600 miners a week leaving the industry for better paid jobs elsewhere, and the miners believed that they were fighting not just for their own levels of pay but for the future of the industry as a whole. The Yom Kippur War which started on 6 October 1973 also brought into focus Britain's dependence upon outside sources of fuel. Whilst this war is often regarded as the main reason for the 1973 oil price shock, the scene for the price rise had been set in April 1973 when the USA removed oil import quotas. This led to a rise in the marginal price of spot oil supplies into the USA, although the average price of the fuel to the US consumer remained little changed. With an increase in demand for tankers, freight rates increased to levels which were normally indicative of a crisis and when the normal freight rates were taken into account, the Organization of Petroleum Exporting Countries (OPEC) was able to calculate an effective price for its oil. This resulted in a unilateral price rise at the start of October, just before the beginning of the conflict. The Arabs imposed an oil embargo on various pro-Israel nations and as there was no response a second, and much larger, price rise was instigated by Iran in December 1973.

Against this background the miners first instituted an overtime ban in order to attempt to reduce stocks of coal around the country. The ban took effect from Monday 12 November 1973. After only 24 hours the government said that it had been forced to declare a state of emergency because of the disruption being caused to the economy. However, part of the reason must have been to attempt to paint the miners in a bad light and sway public opinion behind the government. Nevertheless, the overtime ban was effective and cut supplies of coal to power stations by some 40%.

The reason that this ban was so effective was that miners do not just spend their entire time in the mine digging coal. Much time and effort has to be expended preparing the machines for operation and moving them to the correct area of the pit. Checking for gas and for the security of the roof is also important and much of this work is normally carried out at weekends, and counts as overtime. With the introduction of the ban the work had to be undertaken on Monday mornings and meant that most of the first shifts of the week produced no coal at all. As a result of the disruption which this caused, a large reduction in output was to be expected.

The state of emergency and the overtime ban continued into 1974 when the country was put on to a three day working week in order to attempt to conserve power supplies. At that time there was about 10 weeks' supply of coal at the power stations. These stocks and an easing of the oil supply situation, together with some consideration that the country could move to a four day week, convinced the miners of the need to take more aggressive action. This led to the ballot for a strike which resulted in an 81% vote in favour of taking full industrial action. The strike then started at midnight on 9 February 1974, two days after Edward Heath had called a general election (for 28 February). The main thrust of his campaign was 'Who Governs Britain?' Heath suggested that the strike should be postponed because the miners would not have an effective government with which to negotiate. This, however, would have indicated a politically motivated strike, something Joe Gormley was very keen to avoid and also would have meant that the miners would have ended up negotiating with the winner of the election during the summer. This was something they did not want because of the country's much greater need for coal, and therefore their greater influence, during the winter.

Even the National Coal Board believed that the miners had a good case for a pay rise and the difficulty which the miners had when initially negotiating with the NCB was that it soon became clear that the NCB's hands were tied by the government. Even though the NCB may have wanted to increase the pay of the miners (had it been able to), the government would have prevented this and so the miners soon started negotiating directly with the government. So, after the polls closed on 28 February 1974 and a Labour administration under Harold Wilson obtained a narrow victory the miners, after initial discussions over the Pay Board recommendations, called off the strike which ended on Monday 11 March 1974.

The Pay Board, to which Heath had referred the miners' pay claim, reported on 21 February that the miners were due some 8% more than they had been offered. This news, which came out a full week before the election, must have damaged Heath's standing in the polls. That Heath had no need to call the election can be seen from two facts. First he could have waited for the full report from the Pay Board and second, his term of office had a further 12 months to run. Arthur Scargill recently suggested that 'if he'd waited for the Pay Board he would still have been Prime Minister and we would have avoided the horrors of Mrs Thatcher'.

PLANNING NEVER MADE ANYTHING

Following the Labour victory and the end of the strike Harold Wilson ordered a new Plan for Coal to determine the future of the industry. The problem with this plan was that, as ever, it was written on the basis of the situation at the time rather than being projected into the future. However, we obviously now have the benefit of hindsight and the fear that the Arab nations within OPEC would keep a continual constraint on oil supplies was not correct. This was even despite the problems which radiated out from the second oil shock in 1979.

In 1973/4 the cost of power station coal was about 4.5 p/therm (0.15 p/kWh) compared with some 6–7 p/therm (0.20–0.24 p/kWh) for oil. At the time it was also believed that the world would run out of oil by the turn of the century and that coal would return to being the major and most important source of energy for all industrial economies. That this is clearly not the case is no fault of the planners of the early 1970s; after all some 20 years ago there was very little known about the potential for the discovery of oil in the North Sea or Alaska, whilst further exploration in the Arabian peninsular has found still more reserves of the fuel.

Nevertheless, the planners were right in one respect, and that is that the use of oil as a fuel is limited and that over the longer term its price can be expected to rise. This is also, of course, true of coal and all other types of non-renewable energy. Indeed, eventually, all of the world's power stations could run out of the uranium needed to power nuclear plants. Renewable energy is the great hope for the future but until it has been discovered, the planners of today are unable to factor such a development into their models of how the world will generate its energy requirements in the next century and beyond.

In view of the planners' inability to forecast correctly a command economy, and similarly, nationalization (which involves government control across the whole spectrum of an industry) will be doomed to failure. This is not least because it will stifle any entrepreneurial spirit that could engender a different view of future prospects and prepare for them accordingly. It is also the case that governments should not take the current situation for granted in the belief that it will be maintained into the future, although the population as a whole is just as guilty of this as the international credit boom of the late 1980s and early 1990s has shown. However the current plan to privatize the coal industry through reducing it in size to a level which is operationally profitable may not be the best way for the country to ensure that it has sufficient energy to meet its future requirements (see Part III).

Chapter

6

A sea change in politics (1979–1990)

LABOUR ISN'T WORKING

The General Election of 3 May 1979 saw the Conservative Party returned to govern the country with a majority of 44 seats, a margin not exceeded since Harold Wilson's 96 seat landslide in the 1966 election. The majority of the population had become totally disillusioned with the Labour Party, its infighting and the lack of a coherent economic policy following the 1976 currency crisis when Denis Healey had to go cap in hand to the IMF. This eventually led to the disastrous 1978/9 'winter of discontent' and stimulated the sea change in the country's political outlook that swept the Tories to power. They were led by Margaret Thatcher, who had taken over from Edward Heath in February 1975, following the catastrophic loss of both 1974 elections.

Mrs Thatcher brought to the government a strict monetary philosophy aimed at reducing the role of the 'nanny' state and increasing the level of self-dependence of both individuals and companies. Indeed, in his autobiography covering his years serving under Mrs Thatcher Nigel Lawson quotes from a book by Leo Pliatzky, a civil servant who had worked in the Treasury and at the Department of Trade. He wrote that 'the Conservative government elected in May 1979 was more than just another change of government; in terms of political and economic philosophy it was a revolution.' This revolution was extended to the nationalized industries which were to be privatized and made to stand on their own, with no chance of government subsidies to keep the inefficient ones in operation.

Before privatization there was one big hurdle which the government had to overcome – it had to return the nationalized industries to profit, or at least to show that the existing level of profitability could be maintained, if not increased, under the competitive environment of a free market. The government was lucky in many respects at the turn of events which were to enable it to remain in power for long enough to achieve these aims. This was particularly because by the time of the 1983 general election the country was in such a poor state economically that a Conservative election victory on the back of its economic management would have been unlikely. Indeed, the 1979 Conservative election campaign which featured the slogan 'Labour isn't Working' almost came back to haunt their copywriters because of the continuing high levels of unemployment. This unemployment had been exacerbated by the jolt which the move to free market economics caused to the economy and the problems of the 1981/2 recession from which the country had still not completely recovered. However, the war over the Falkland Islands led to a massive feeling of patriotism and this enabled the Conservatives to win an overwhelming 141 seat victory. This was an event which Arthur Scargill declared as the 'worst national disaster for a hundred years'.

That there was going to be a showdown with the miners at some stage was indisputable, especially following the link which so many people in the Conservative Party place between their loss of the first 1974 election and the miners' strike which had started 21 days before. In 1982 Margaret Thatcher, a 'lady not for turning' had made an amazing U-turn over her policy of returning the mines to profitability through a massive pit closure programme (one which would have been much more severe than that announced on 13 October 1992). This humiliation, probably more than the loss of the 1974 election, put her on a confrontation course with the newly elected President of the National Union of Mineworkers, Arthur Scargill, in 1983.

That Mrs Thatcher was able to confront the miners was undoubtedly related to two factors. The first was simply the massive closure programme which had gone before. The number of mines and miners had diminished slowly but surely and meant that there were many fewer in the industry who could cause the massive disruption to energy supplies which had been seen in the strikes of both 1972 and 1974. The second advantage which Mrs Thatcher had over her predecessors was that a whole raft of new trade union legislation had been introduced which disallowed secondary picketing and also disallowed sympathy action among related unions which could cause further energy disruption. In effect the old triple alliance of the miners, railwaymen and the transport workers had been smashed and the need for mass picketing, whilst used in the 1972 and 1974 strikes, became greater in order to ensure effective disruption of the supply of coal.

That the government was expecting a strike there can be no doubt as it spent some time preparing itself for that eventuality, both through the appointment of Ian MacGregor as the Chairman of the then National Coal Board but also with the build-up of coal at the power stations and pit-heads. This would increase the length of time for which the country could continue with few adverse effects from a strike and thereby reduce its effectiveness in achieving the miners' demands. According to Nigel Lawson, this was not supposed to be an aggressive move but rather a deterrent as the knowledge of high stocks of coal around the country was supposed to put the miners off striking as it meant that the country would not suffer unduly from a protracted battle. Indeed, Nigel Lawson has stated that Mrs Thatcher's first words to him on his appointment as Minister of Energy in September 1981 were: 'Nigel, we mustn't have a coal strike'. However, Lawson was under no illusions about Arthur Scargill and says that 'when Scargill was appointed it was clear to me that he was going to go for a strike' and his policy was therefore aimed at countering that probability.

SCARGILL V THATCHER

The 1984/5 miners' strike was the longest and most aggressive strike that has ever taken place in this country. The roll call is as follows: two working miners died and 255 were injured as a result of picket line violence; two working miners committed suicide; and 790 police officers were injured, two of them seriously. The financial cost of the strike is a contentious issue; however, in a supplementary note to the House of Commons Civil Service Treasury Committee, Appendix 7 to the second report in the 1985/6 session the Treasury admitted that it cost the country £3.75 billion. Alternatively, in his autobiography Nigel Lawson records that 'the public expenditure planning total was exceeded by over £3.5 billion in 1984–85, two thirds of the overspend being attributable to the strike. In the same financial year the strike added £2.75 billion to the Public Sector Borrowing Requirement' (p 160). The strike was about many things – a struggle between left and right and a struggle for the security of jobs for the miners in the long term. Specifically, however, the strike was based on Arthur Scargill's belief that no pit should be shut solely for economic reasons. A closure on the grounds of safety or exhaustion would be permitted but that the NCB was not making any money from a pit was no reason to close it. In some respects such an assumption is perfectly fair and reasonable but, and this condition is sacrosanct, this can only be the case if there is a chance for improvements either in productivity or in price which would return the mine to profit in the future, such profit exceeding the losses and interest costs which would have to be sustained over the intervening period. Arthur Scargill's failure to accept this condition, and Ian MacGregor's insistence on it, led to the strike continuing for a full year.

In the strike Arthur Scargill made one great mistake and this was an error which caused him to lose support, even before the dispute had started in earnest. The reason for the error was simple – it was that Scargill was unsure that he would be able to obtain the necessary majority in a ballot of NUM members to call a strike. Indeed, he had lost three ballots in the first 18 months of his tenure of the NUM Presidency. He therefore utilized the NUM rule book which allowed a strike to be called on compliance with one of two conditions. These were Rules 41 and 43. Rule 43 states that a 'national strike shall only be entered on as the result of a ballot vote of the members' and under NUM rules 55% of the votes had to be cast in favour of strike action. Rule 41 covers local area strikes which are allowed to proceed only after appproval has been sought and obtained from the NUM executive committee. As a result of Rule 41, Scargill believed that he had the capability to call a strike in one area and then expect all other areas to walk out in sympathy (especially if they were encouraged to do so by picketing at the colliery gate). Therefore he did not call a full and proper strike ballot before ordering that the industrial action should start.

The sequence of events started in November 1983 when an overtime ban in rejection of the NCB's 5.2% pay offer was instituted. Then in late 1983 Mick McGahey, then Vice-President of the NUM, met the Deputy Chairman of the NCB, Jimmy Cowan, and advised him that, for whatever reason, a strike was likely to start in March of the following year. Whether or not the NCB played into this timing, because they could see the stupidity of calling a strike over the summer when the overtime ban had failed to reduce stocks at power stations, or whether it was the natural sequence of events is not known. However, on 1 March 1984 George Hayes, the South Yorkshire area director, announced the intention to shut the Cortonwood pit. It had costs of £64/t and had lost £4 million in the previous year; the NCB had every reason to decide that the pit was no longer economically viable. On Monday 5 March the South Yorkshire NUM took a decision to strike from 9 March, whilst at a meeting between the NCB and the NUM on 6 March, under the aegis of the Coal Industry National Consultative Council, the Coal Board's plan to cut 4 mt of capacity was outlined.

The NUM executive met to approve the South Yorkshire strike on 8 March and considered two motions, the first allowing a Rule 41 strike and the second calling for a Rule 43 ballot. The first motion was carried by a 21-3 majority but the second motion, which may also have been passed, was never put. Initially many local areas, whilst acknowledging the problems of the closure of the Cortonwood colliery, believed that there should have been a national ballot and organized their own local votes to determine the mood of their members. In South Wales 19 pits voted to work and 11 to strike but by the start of Tuesday 13 March all of the mines had been picketed to a standstill. In North Derbyshire the abuse of democracy was even more startling. The area organized a ballot on Friday 16 March which resulted in a vote of 4307 for a strike, but 4323 against. However, the NUM executive (assuming that they would have voted against strike action) disallowed the votes cast by a local NCB laboratory and a non-NCB mine, thereby engineering the 55% majority required for strike action. As a result of these actions and of the effectiveness of the picketing, only 11 of the NCB's 191 collieries were working by Saturday 10 March.

After this there was a long period of conflict, with little movement on either side. The first meeting between the NUM and the NCB since the start of the strike was on 23 May, the start of the month long 'battle' for the Oregreave coke works. However, this and all subsequent meetings between Arthur Scargill and Ian MacGregor came back to what MacGregor called his right to manage the business and Scargill's refusal to concede pit closures on the grounds of economics alone. This, ultimately, led to Scargill's downfall, as his intransigence and inability to negotiate infuriated not just the NCB but also members of the TUC who were also attempting to resolve the dispute. This is in marked contrast to Joe Gormley about whom it was written, 'Joe could bargain the buttons of their [the NCB's] trousers' and who proudly acknowledged, 'negotiating has been my life' (*Battered Cherub*, p 182).

Probably the most harrowing part of the dispute came in September when the National Association of Colliery Overmen, Deputies and Shotfirers (NACODS) threatened industrial action. This would have halted output at those pits which were still producing coal because of the lack of sufficient safety cover. Indeed, Roy Lynk, the then President of the UDM, stated to the Trade and Industry Select Committee on 2 December 1992 that, even today, 'if there is a dispute between NACODS and British Coal then the UDM men get sent home with no pay' so important is the role of the NACODS man in the pit.

The NACODS dispute centred on the different rules about turning up for work that applied to NACODS members during the NUM strike, which depended on the NCB area in which they worked. They considered this unsatisfactory and at a delegate conference on 12 September decided on a ballot for strike action to be held at the end of the month. The result of the ballot was 82.5% in favour of strike action and it was set to start on 25 October 1984. By agreeing, at the last minute on 23 October, to allow an independent review to be included in the Colliery Review Procedure for the closure of pits the NCB averted a strike. However, this may only have been because NACODS misunderstood the power of the independent review which could only argue against, and not prevent, pit closures.

In November 1984 the NCB instituted a massive publicity campaign, pointing out that miners could earn £1400 by Christmas if they returned to work. This campaign was a great success and miners flooded back to work. On 8 November the first one went back at the Cortonwood pit, at the centre of the dispute, and others soon followed him. By Ian MacGregor's reckoning the dispute would have finished 'when 51% of our people are back at work'. This occurred on Thursday 28 February 1985, and by 5 March the strike had ended. It was a long and bitter battle – particularly for those miners who chose to work rather than be forced into an unnecessary confrontation which was going to cost more closures and more jobs than the strike itself was attempting to prevent. Some 12 000 miners left the industry during the dispute whilst the irreparable damage caused to the collieries by a year of lying idle caused nearly 40 mines to close and 38 000 people to lose their jobs.

One of the results of the dispute was the break of the Nottinghamshire miners from the NUM to form the Union of Democratic Mineworkers (UDM). The path which led to the eventual formation of the UDM started in May 1984 when a number of working miners were upset by the lack of democracy which existed within the workings of the NUM. A campaign in Nottinghamshire and South Derbyshire continued throughout the dispute and the Nottinghamshire branch of the NUM was eventually expelled from the union at a meeting on 10 January 1985. This led directly to the formation of the UDM some six months later, under the Presidency of Roy Lynk, formerly number 3 in the Nottinghamshire NUM hierarchy.

Following the announcement of extensive pit closures in October 1992 many members of the UDM (whose pits were as badly hit by the closures as those worked in by the NUM) consider that Arthur Scargill was right in his main belief that the Conservative Party wanted to decimate the industry. This led to the November 1992 removal of Roy Lynk from office, although he is still campaigning to get British Coal to sell or lease some of the pits which it would otherwise close. Many people across the country were firmly of the belief that Arthur Scargill was a rabid communist whose only aim was to bring down the government in the same way that the 1974 strike was supposed to have brought down the Heath government. That it was his belief that the Conservative government intended to close the bulk of the British coal mining industry and that therefore the government had to be opposed may be seen to be a rational argument in the light of subsequent events. However, it excludes the need for a profitable industry which is the only way to guarantee long term job security. It was this precept, more than anything else, that meant that the government and the NCB refused to give in to Arthur Scargill as a government cannot continue to subsidize a loss making industry forever.

Unfortunately for Arthur Scargill, and indeed the jobs of the miners which have been lost over the years since the strike ended, like any good general Margaret Thatcher had planned her campaign carefully. The first and probably most important move was to put the mines under the control of someone who would not be frightened or overawed by the task in front of him. Ian MacGregor, who had been appointed to run British Steel by the then Secretary of State at the Department of Industry, Keith Joseph, had brought about a dramatic turnaround in the fortunes of the company (although it also reduced that company's demand for coal in the process). Joseph initially suggested MacGregor to Nigel Lawson but the continuation of MacGregor's contract at British Steel prevented him from taking the position full time, and initially, Norman Siddall, the deputy chairman was appointed in place of Derek Ezra, the previous chairman.

Whilst MacGregor was keen to take the job on a part-time basis Lawson was convinced that it would be a full time occupation and that MacGregor would not be able to run two nationalized industries at the same time. He also had to take into account the massive political implications that such a dual appointment would raise. Eventually he secured agreement from his Cabinet colleagues for the switch and persuaded MacGregor to leave British Steel and take the job in February 1983. An earlier leak meant that the March 1983 confirmation of MacGregor's appointment did not cause as much of a furore as initially anticipated although he soon earned the nickname 'the geriatric American butcher'. However, when he joined the NCB in September 1983 he made great play of the fact that he was born and brought up in Scotland and that, as far as he was concerned, he was not geriatric!

Unfortunately, with the miners continuing in employment and working the overtime necessary to maintain production, thereby ensuring that pits remained open, stocks of coal at the power stations and other depots around the country were increased. Some power stations were also converted to be dual firing rather than just coal fired; in his autobiography Nigel Lawson states that Mrs Thatcher wanted every coal-fired power station to be converted to dual firing, and hang the expense. Nevertheless, by the time the strike started, the country had stockpiles of over 60 mt of coal and was less dependent upon coal as a source of energy than it had been previously.

The other advantage Mrs Thatcher had over Edward Heath was that during the 1970s the country had been developing its vast oil and natural gas reserves in the North Sea and at various sites on the mainland. These reserves meant that Britain was much less dependent upon imports of oil and gas than it had been previously and that a sustained rise in the price of oil, like the one following the Yom Kippur War, would have a less serious effect on the country's balance of payments. Gas had first been discovered in the North Sea in September 1965 and even by 1970 it had started to take some 25 mt of demand away from coal. This was in the domestic market where gas was used in cooking and heating systems and although the old coal gas systems had to be altered to enable them to use natural gas the progress of natural gas across the country continued until the entire mains supply had been converted to run on natural gas by September 1977.

Having laid the ground plans so carefully it was no surprise that the miners were forced to capitulate to the NCB's demands and the strike ended. It had a damaging effect on both the unions and the coal industry and, as already mentioned, led to early closure of some mines as they became geologically unsafe. The other area of industry which became disillusioned with coal was the Central Electricity Generating Board (CEGB). This energy supplier had previously relied on coal for a large amount of its supplies but the return of cheap oil, and hence a fall in international gas prices, heralded a new era of oil- and gas-fired electricity generation. This has had more recent effects on the electricity generation industry due to the desire of the managers who suffered during the 1984/5 strike to move away from their dependence on coal as a main component of their generating capacity. Indeed, as Ian MacGregor pointed out to the NUM,

> electricity generated from coal accounted for 80% of the CEGB output and purchasing it amounted to half their costs. . . . this comfortable arrangement would clearly not go on for ever and, unless we did something about our costs, other sources of electric power – nuclear and even oil and gas – would have a growing impact.

How right he was.

PRIVATIZE EVERYTHING

Minimal state intervention was one of the main philosophies of Thatcherite Conservatism. Added impetus was put into a reduction in the amount of red tape, in the form of unnecessary or obstructive regulation, with which the population had to comply. Privatization was one of the methods which the government could see would help it to achieve this ambition. This is because a privatized industry is more interested in generating profits than in collecting and collating statistics or having forms completed in triplicate, as these activities will detract from its earnings potential. The government could also see that, whilst it would mean an increase in unemployment because those workers surplus to requirements would be made redundant, it would be likely to lead to higher growth as companies reinvested their increased earnings in new plant and equipment.

Unfortunately, as with the Labour Party's nationalization programme in the late 1940s, the Conservative Party's programme was affected by political constraints. This meant that it was sometimes driven by ideology rather than by sound commercial judgement and that it did not necessarily obtain the best possible price for its sales. In one respect, of course, the Conservative Party was engaged in trail-blazing and the rest of the world soon started to follow suit earning the UK substantial advisory fees. However, there were some massive upsets both with the undervaluation of the Amersham issue and the problems associated with the collapse in the oil price in the run-up to the sale of the first tranche of Britoil shares. Nevertheless, there have been fewer upsets recently, although the general malaise of the stock market following the October 1987 crash has led to a reduction in interest and hence post-issue premiums.

Overall a total of 46 companies with some 900 000 employees have been returned to the private sector where, there is little question, they should always have been in the first place. The size and breadth of government intervention in industry was staggering and, of course, it meant an expansion in the number of government departments necessary to oversee all of the activities of state controlled companies (as an aside the privatization of all of Britain's energy suppliers bar British Coal, the nuclear generators and the Northern Ireland electricity company led to the abolition of the Department of Energy after the April 1992 election).

However, there was a further plank to the Thatcherite philosophy that required the privatization programme to continue, sometimes at inauspicious moments. Probably the best remembered of these was the government's decision to proceed with the sale of a further tranche of BP shares following the 1987 crash. The reason for this determination, though, was that the government needed privatization proceeds to balance its books. This was because, as a further incentive to private industry, the government was determined to reduce the level of direct personal taxation so that people could determine where to spend a large proportion of their earnings. The expectation was that this would have two effects; first people would feel better off and second, as a result of this, they would have a greater incentive to work harder and earn more money, thereby covering the shortfall that would be left by the absence of privatization receipts over the longer term.

Unfortunately the trouble arises for a government which commits itself to too much spending and then does not have the wherewithal to finance it except from the continuation of the privatization programme. This meant that the government was forced into many of the later privatizations at times to suit its funding requirements rather than taking the time to ensure that it obtained as much as possible from the sales. The programme is not complete and so there is little likelihood that there will be a let up in new issues despite the poor prices that could be obtained by selling industries in the current recessionary environment (note that the projected PSBR for 1993/4 at £50 billion now represents a massive 8% of forecast GDP against repayments in the late eighties). This negates some of the fiscal prudence arguments of the 1980s when the government was supposed to have reduced borrowings to such a low level that it would not be difficult to fund the increased spending needed during a recession. This is so clearly not the case in 1993, although there is some excuse to be gained from the extreme length of the downturn, that the success of the government's monetary policies under Margaret Thatcher must be brought in question.

Indeed, it was these policies which meant that the government had to push ahead with the privatization of the electricity industry in the three separate tranches of the regional electricity companies in England and Wales, the two generators and finally the two Scottish generating and distribution companies before the privatization of the coal industry on which the generators are dependent. This was a complete reversal of the natural progression of privatization, which should have started at the base rather than the fragmented top of the network. As a result, coal has now had to be squeezed to fit into the existing structure rather than forming the foundation on which the electricity industry can grow.

Unfortunately the coal industry did not see this as much of a problem at the time and the anti-nuclear lobby had a much more concerted campaign which forced the government to abandon its plans to include the country's nuclear generating capacity in the disposal. This has meant that the nuclear industry has remained under government control and that it receives a subsidy which means that there was an additional 11% (now reduced to 10% for 1993) added to the cost of electricity generated from fossil fuels.

GETTING GREENER ALL THE TIME

Another of Mrs Thatcher's legacies to the country was her increasing environmental awareness. It was clear to her, as a scientist, that there was a massive imbalance between the production and consumption of greenhouse gases in the world because of the destruction of tropical rainforest. This led to a number of summits at which she spoke about the need for the developed countries of the world to reduce, or at worst stabilize, carbon dioxide emissions from power stations and other carbon fuel consumers. A vast amount of disagreement sprang up between the developed and the developing world about who should pay the cost of installation of new, greener equipment. The other problem which arose was that the developing countries refused to accept that they should suffer because of all the pollution which had been churned out by the developed countries as they went through their own industrial revolutions in the nineteenth and early twentieth centuries.

Nevertheless, the British government agreed that it would comply with European Community legislation to limit the emission of carbon dioxide and other greenhouse gases in order to attempt to reduce the deleterious effects of global warming. Specifically it is required under the Rio Earth Summit that carbon dioxide output is stabilized at 1990 levels by the year 2000, a restriction that has had a knock-on effect on the electricity supply industry. The European Community legislation requires the UK to reduce its sulphur dioxide emissions by 20%, 40% and 60% by 1993, 1998 and 2003 respectively. These regulations are less onerous than those for France or Germany where the 40% target has to be reached earlier and the reduction by 2003 is 70%. Nevertheless, the electricity generators have been forced to install new and more efficient exhaust gas purifiers in order to attempt to remain within these guidelines and they have also been encouraged to move away from the dirtier coal- and oil-fired power stations to more environmentally friendly production of electricity. This, obviously, has had an effect on the demand for coal from the electricity generators, although the 1990 EC directive which permitted the generation of electricity from natural gas for the first time has, obviously, also had an effect.

CLOSING DOWN THE MINES

That the closure of the coal mines continued, almost unabated, cannot be disputed and makes a mockery of the government's supposed new for old policy. The miners attempted to make waves and to show the country what was happening but they were not only split as a result of the breakaway of the UDM but also thoroughly demoralized due to the length and bitterness of the 1984/5 strike. There was no way that they had the where-withal to strike again, just as after the General Strike in 1926 the miners were

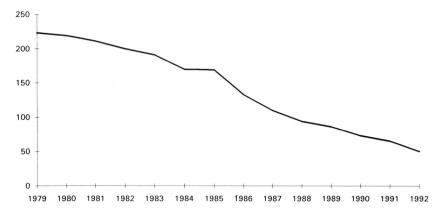

6.1 Number of NCB/British Coal collieries 1979–92

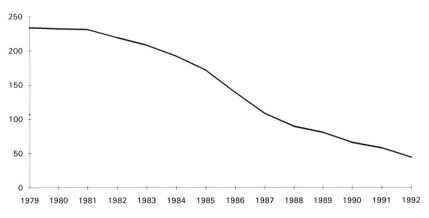

6.2 Number of miners on colliery books 1979–92 (000s)

unable to gather themselves together to face a further confrontation until 1935.

The other problem which the miners faced was that as a result of the violence of the 1984/5 strike they had discredited themselves amongst the general population of the country. As a result they were dismissed as members of the 'loony left' and the sense or otherwise of their argument was ignored rather than even disagreed with. The previous strikes, as the first 1974 election showed in graphic detail, were widely supported by people who could see that the miners had been very badly treated by successive governments, not just Conservative ones.

Under the Thatcher administration Britain's coal industry was transformed. This transformation occurred at the cost of mines (Figure 6.1) and jobs (Figure 6.2) as productivity improvements, which were made possible by the lifting of union intransigence following the 1984/5 strike, meant that fewer workers could produce the same amount of coal. At the end of the day privatization was, and still is, the ultimate goal of the Conservative Party and the pursuit of this aim during the Thatcher years led to a marked improvement in British Coal's international standing (it changed its name from the National Coal Board in 1986). Nevertheless, the task was incomplete when Mrs Thatcher was unceremoniously dumped by the Conservative Party in November 1990 and it was up to her successor, John Major, to attempt to finish the task that she had started (see Chapter 7).

Chapter

7

The conductor changes (1990–1993)

AND THE BAND PLAYED ON

Whilst the election of John Major as leader of the Conservative Party, and Prime Minister, in November 1990 occurred in the knowledge that he was Mrs Thatcher's chosen successor it soon became clear that he was not her dummy. And, when he obtained his own mandate from the electorate in the Conservative Party's surprise April 1992 election victory he appears to have been less threatened by her legacy. However, many of the basic tenets of Thatcherite philosophy remained unaltered and the so-called privatization bandwagon rumbled on.

Nevertheless, in his 'Vision of Britain' the new Prime Minister was very careful to portray a more compassionate nature than that of his predecessor. This, if anything, left him more exposed to the fluctuations of political fortune and meant that it was much harder to push through and justify unpleasant policies. Such was the situation in the fiasco of Black Wednesday on 16 September 1992 when the previous indecision either to raise interest rates or revalue sterling within the Exchange Rate Mechanism (ERM) came to a head. This inflexibility over the value of the pound forced the government into an embarrassing defeat at the hands of the very market of which it approved.

As Norman Willis, the leader of the Trades Union Congress, stated when complaining about the pit closure programme 'the government seemed frozen like a damned rabbit in the headlights of the recession'.

In the realization of when it is necessary to capitulate to the demands of others and when to pursue their chosen path the two leaders could not be more different. Part of this may be related to John Major's relative inexperience in government although he has learnt a lot in his first few years in office. But whatever their psychological differences the ideals of the free market were maintained. As a result Cecil Parkinson's 'ultimate privatization' remained firmly entrenched in Conservative Party beliefs – even if it meant that a large number of the remaining collieries still had to be closed.

BLEAK OCTOBER

On Tuesday 13 October 1992 British Coal had 50 collieries in operation, with a further one, at Asfordby, under construction and due to come to production towards the middle of the decade.

In the two financial years preceding the closures British Coal had made net profits of £78m and £170m respectively (see Chapter 17). That the company appeared profitable led the public, and many Conservative Members of Parliament, to treat the decision with contempt. However, the profits which British Coal had earned came from its position as an effective monopoly supplier to the electricity generating industry. Unfortunately for British Coal it had taken too long to adapt to the financial rigours of the free market and, as a result, many of its pits would be unprofitable at the much lower prices at which new supply contracts were to be negotiated.

The decision to close pits on economic grounds should have been no surprise to the country. It was clearly foreshadowed in the October 1991 leaks of a report by NM Rothschild, the merchant bank advising the government on the privatization of the industry. The bank forecast that only 14 of British Coal's pits, employing 11 000 miners, could be privatized, although only 12 pits were likely to be in operation by 1995. At the time of the leak John Wakeham (now Lord Wakeham) the Minister of Energy stated that 'the future size of British Coal's operations will depend on the size of the UK coal market and the share of that market that they can win'.

This was a clear pronouncement of the government's intention that British Coal should become a profitable business within the constraints of a free market in the supply of energy. There should have been no doubt inside the company that this was the government's policy – indeed it had maintained the same stance towards the coal industry, despite a change in leader, since the 1979 General Election. Perhaps British Coal had expected that the Labour Party would be swept to power in the 1992 General Election and that there was therefore no need for concern because the new administration would force the generators to buy British-produced coal whatever the price. If this was the case they were wrong and the continuing high cost of production across the industry meant that pits were consigned to the closure list because they could not compete with foreign competition.

However, as 1992 progressed the succession of newspaper reports and leaks gathered momentum and it became clear that a major announcement on the future of the coal industry was imminent. This was apparent on 17 September 1992 when Arthur Scargill released a leaked letter from Tim Sainsbury, an Industry Minister, to Michael Portillo, Chief Secretary to the Treasury. It detailed British Coal's intention to close up to 30 collieries with the loss of more than 25 000 jobs but received little press coverage at the time. The main reason for this was that the front pages were filled with the financial dramas of Black Wednesday, but it was also the case that Arthur Scargill had cried wolf about pit closures too many times before.

Then, during the course of the morning of Tuesday 13 October the population woke up to the fact that Arthur Scargill had not been making mountains out of coal stocks – indeed the decisions proved even more precipitous than the fears that Scargill had previously expressed. The plan, detailed in the Table 7.1, was that British Coal would close a total of 31 pits, 27 permanently, with four to be placed on care and maintenance. This would enable the four pits to be reopened if the demand for British-produced coal grew and further coal production could be justified. A total of 30 000 jobs would be lost in the mining industry directly, and there was expected to be a large number of jobs lost in related industries. Estimates of up to a total of 110 000 job losses have since been published which showed the seriousness of the situation – particularly because of the disastrous state of the British economy at the time. These forecasts, however, also included estimates of the knock-on effects of the disintegration of areas in which, in some instances, the number of lost jobs would double as a direct result of the closures. Of these the loss of jobs in the railway industry was expected to be the next largest at some 5000, as there would be less coal to transport by rail and British Coal had been one of BR Freight's largest customers. At one stage National Power was contemplating the acquisition of British Rail's freight operations.

When the announcement was made many politicians on all sides stated their total abhorrence of the apparently uncaring way in which the miners had been treated by the government, including many members of the Conservative Party who also spoke out against the plan. In particular Elizabeth Peacock, MP for Battle and Spen, and Winston Churchill, MP for Davyhulme, announced

Table 7.1 Pits earmarked for closure in October 1992

Good pits	Marginal pits	Closure pits
Annesley/Bentnick	Hatfield/Thorne	Bentley
Asfordby (open 1995)	Maltby	Betws Drift*
Daw Mill	Prince of Wales	Bevercotes
Ellington/Lynemouth	Wearmouth	Bilsthorpe
Goldthorpe		Bolsover
Harworth		Calverton
Kellingley		Clipstone
Littleton		Cotgrave*
Longannet		Easington
Manton		Frickley
North Selby		Grimethorpe*
Ollerton		Houghton Main*
Riccall		Kiveton
Silverwood		Markham
Stillingfleet		Markham Main*
Thoresby		Parkside*
Tower		Point of Ayr
Welbeck		Rossington
Whitemoor		Rufford
Wistow		Sharlston
		Shirebrook
		Silverdale
		Silverhill*
		Taff Merthyr*
		Trentham*
		Vane Tempest*
		Westoe

* 10 pits excluded from review (see p 93)

their intention to join Roy Lynk, the Chairman of the UDM, at a sit-in in his former colliery, Silverhill. They were eventually thwarted in their attempts to go down the mine but continued to press the government to change its mind over the closures. Dr Michael Clark, MP for Rochford and a long standing member of the defunct House of Commons Energy Committee, also spoke out vehemently against the closure programme and continued his pressure through his appointment to serve on the Trade and Industry Committee. Whilst Winston Churchill was swayed by the announcement of the review and voted with the government following the debate on the pit closures on Wednesday 21 October, both Michael Clark and Elizabeth Peacock voted against. Elizabeth Peacock was sacked from her position as Parliamentary Private Secretary (PPS) to Nicholas Scott, the Social Security Minister, following the vote.

The problem was that the closures were, in many ways, inevitable following the privatization of the electricity industry and the subsequent desire of all the regional electricity distribution companies (RECs) to reduce their dependence on PowerGen and National Power for their electricity supplies. When building a new power station from scratch it is cheaper to build a gas-fired station than it is to build a coal-fired station and therefore all of the electricity distribution companies opted for gas. This meant that the two big generating companies had excess generating capacity which was not required, and are now being forced to close power stations down. It seems strange that none of the distribution companies thought about purchasing some of the power stations which were being closed rather than having to build brand new ones from scratch. However, when the leash of government control was slipped, all the RECs chased after the same fox rather than turn back to the bowl which had given them succour in the past.

A government inquiry under the auspices of the Office of Electricity Regulation (Offer) had been commissioned by the Offer Chairman, Professor Stephen Littlechild, in the autumn of 1992. The remit of the inquiry was to consider whether or not the electricity companies were staying within their guidelines, and to ensure that they were paying the minimum amount possible for their electricity, in order to confirm that the consumer was not being forced to pay more for electricity than is fair and reasonable. The results of the inquiry, conducted by a company called London Economics, were originally due to be received by Offer in January 1993 and it was then to be considered by Offer which would publish its conclusions during the summer of 1993. Public outcry following the announcement of the closures forced Offer to speed up the proposed investigation so that it could present its report early in 1993. This ensured that its conclusions were available at the same time as the results of the investigations conducted by the Trade and Industry Select Committee, the DTI (which itself hired a number of consultancy firms) and Boyd's which had been asked by the DTI to take a second look at the structure and costs of the industry.

The reason for Offer's investigation was the widespread belief that the use of coal in power stations, even if it is more expensive British coal, is still cheaper than using gas because of the capital costs of constructing the gas power stations (£300 million to £400 million, about the same as a new coal mine). One of the arguments which was put forward was that coal stations are not economic or environmentally friendly because of the vast amounts of sulphur and carbon dioxide they emit into the atmosphere. Nevertheless, whilst it is not possible to reduce the emission of carbon dioxide, it is possible to reduce the amount of sulphur which is released through the installation of flue gas desulphurization equipment (FGD). This equipment is widely used in the USA and in Germany and has been shown to be effective. The cost of installing such equipment into British power stations is estimated to add about 0.5p/KWh to the cost of coal-fired electricity.

One of the main complaints of the Conservatives, which was repeated by Michael Heseltine, President of the Board of Trade, time and again in interviews, was that the miners had refused to modernize and that they retained restrictive practices across the industry as a whole. The industry, however, had also sucked up some £18 billion of government funds during the Conservative administration between 1979 and 1992 and much of this money had been spent on new equipment in order to increase output and productivity, although much was also used to make redundancy payments as mines were closed. Total British Coal production in 1979 was 109.3 million tonnes and the average production/employee was 470 tonnes/year. The figures for 1992 were 71 million tonnes and 1357 tonnes/year respectively, showing that all the mines were increasing their productivity and that they were becoming much more competitive in the market.

The most obvious example of this must be seen in the reduction in the price of coal which the electricity companies are likely to pay in 1993 and the further reductions which British Coal expects over the following four years. This also compares with the massive decline in the cost of electricity generated from coal between 1974 and 1993. That the government was willing to invest in the industry and then close it down before the full benefits of the investment were allowed a chance to take effect smacks of incompetence as the investment in mines destined for closure was clearly a waste of public money. It also suggests that the government may have been trying to lull the miners into a false sense of security by making them believe that the rumours about pit closures had no foundation. And, although it would have been British Coal that decided where to spend the money it must be remembered that a confrontation with the miners ahead of the 1992 General Election would not have been favourable to the government. However, the government's need to raise privatization money to reduce the PSBR must also have played a part in the decision to press forward with the closures. The final tranche of British Telecom shares, which will be sold in 1993, confirms this as does the fear expressed by some commentators that Oftel's bite will only then start to draw more blood from BT's effective monopoly. Indeed, the short term management of a long term asset is the major gripe the government has over the financial markets of the City. Yet if the government proceeds along the same lines can the City be blamed for changing its own expectations accordingly?

That the Cabinet had not discussed the pit closures in full session, but had left it to a subcommittee, is peculiar. The Major style of government was supposed to be that the whole Cabinet should discuss matters of such national importance as the decimation of the mining industry and the increasing dependence of the electricity generators on oil, gas, nuclear and hydro-generation plants. During the exchange rate problems which absorbed much government time in September, the government promoted the fact that the decision to come out of the ERM was made by a meeting if not of the full Cabinet then certainly of the most important figures within it.

Michael Heseltine, however, has been thrown deep into the fray. In his speech at the Conservative Party conference in 1992, when he knew that as a result of his desire not to intervene some 30 000 miners would lose their jobs, he stated that he would 'intervene before breakfast, before lunch and before dinner' to help British business. His argument about the pit closures was that he was, indeed, intervening and that his intervention had led to a £1 billion redundancy package for the men who were being made redundant. This intervention, he also contested, was necessary because the country would otherwise be forced to continue paying too much for its electricity due to the higher cost of electricity generated from British-produced coal as opposed to imported coal or gas. This statement seems, clearly, to be inaccurate to say the least (see Chapter 13), and must be the result of the government being given bad advice by its civil servants or, to quote Richard Caborn, the Chairman of the Trade and Industry Select Committee, of its 'war of attrition with the miners'.

Indeed, some of the anti-mining comments from the right wing of the Conservative Party must have been uncalled for in the light of the pain and anguish which so many people and their families are to suffer. The backlash which these comments generated is unsurprising, although the backlash from the backbench members of the Conservative Party itself is highly unusual. Two Conservative backbench MPs, Winston Churchill and Elizabeth Peacock, have already been mentioned and even Sir Marcus Fox, the Chairman of the 1922 Committee, stated that 'a review is imperative' although he would not vote against the government. However, he implied that he would only vote with it if there were some changes to the programme of closures which he was hoping to obtain in behind the scenes negotiation.

One further astounding facet of the pit closures was that the government and British Coal had shown scant regard for the rules and regulations govern-ing pit closures which they had themselves introduced. In December 1984 British Coal had told the miners that there would be no compulsory redun-dancies and yet the closures which were being announced were all compulsory. Secondly, the closures were all taking place without any consultation with the unions and employees. Under both UK and EC law employees have to be included in any discussion to close the coal mines and there is a recognized Colliery Review Procedure which can be invoked before a mine can be closed. This is probably the reason why Roy Lynk said he was 'gobsmacked' when he heard the news of the closures, not just because of the backing he effectively gave the government during the 1984/5 strike, but because of this ignoring of the legal process.

On the day that the first four collieries were due to close, Friday 16 October, a High Court injunction was issued which prevented the closures from taking place for at least a week. The judge in the decision said that he would issue a statement on Tuesday 20 October and that he was unable to do it as planned on the Friday. The delay of the judgment, which meant that the NUM had only temporarily obtained a stay of execution, also meant that there could be a full debate on the pit closures in Parliament which reconvened from the summer recess on Monday 19 October when a statement was due to be made. On the following Wednesday, the 21st, a full debate took place.

By the time of the debate the government had changed its position to the extent that 21 of the 31 pits on the original list would remain open whilst the DTI completed a full review of the industry, with the help of British Coal and its American mining consultants, John T Boyd (see Table 7.1). The Boyd's investigation was concerned with the profitability of the 21 pits and ranked them in terms of their future potential, whilst it also studied the 19 pits which were to remain open come what may, in order to ensure that this should be the case. The 10 pits which were not saved from closure were just subject to the statutory 90 day consultation period which covered employee problems. This did not constitute a thorough review under the aegis of the Colliery Review Procedure and, as no information on the future potential of the pits was made available, it was the basis of Arthur Scargill's claim that British Coal had something to hide. And, in many respects it may well have had something to hide because the first Boyd's report in April 1992 confirmed that 27 of the 28 pits the consultants were asked to review would be able to produce coal at an average price of £1.33/GJ by 1995/6.

Although Boyd's did not continue to investigate the potential of Betws Drift following their site visit, as they considered the level of reserves was too small, the 28 pits covered in their report included five of the 10 pits earmarked for definite closure (Betws Drift, Grimethorpe, Houghton Main, Parkside and Trentham). That Betws Drift should be excluded from the review due to the short life of its reserves shows the difference in opinion of the American coal mining industry and the UK industry. Michael Clapham MP admitted the short life of the reserves but pointed out the opportunities for a smaller scale and longer life operation, and the potential for an increase in the revenue per tonne of the colliery, something which was not considered by

Boyd's because of the large scale and different operating environment of the American industry in which they have their expertise. The problems were particularly apparent when the consultants stated that they did not look into the market for the coal but just considered each mine on its mining merits. If Betws Drift could increase its revenues from £70/t to the £200/t suggested by Michael Clapham, then the mine may well have been viable. Boyd's effectively signed its death warrant despite not investigating the full facts of the operation.

Whilst the choice of the 28 pits to be investigated by Boyd's was up to British Coal, that five of the mines should be slated for definite closure so soon afterwards shows a serious lack of judgement somewhere within the British Coal hierarchy. However, as Ronald Lewis, the witness for Boyd's who gave evidence to the Trade and Industry Select Committee said, 'British Coal culture reflects an industry which has operated in a national setting for 45 years. . . . much of the attitudes are somewhat constrained' and many of the operations 'would follow a headquarters' line' with little local incentive or flair.

THE PROFITABLE DECLINE

Of course it must not be forgotten that the pit closures announced on Tuesday 13 October 1992 were not the whole story. When the Conservatives came to power in 1979 there were 223 collieries producing deep mined coal under National Coal Board (as it was then) ownership. At the time of the pit closure announcement there were only 50. Over the intervening 12 years there had been a massive decline in the industry to attempt to shrink it to profitability. As mentioned above, this goal had been achieved but, and this was the problem, only because the company was operating in a false market and receiving inflated prices for its output.

The actual number of the closures, however, is not the issue which intensified the horror and the surprise of the general population – after all it was the threat of a similar number of closures which brought about Mrs Thatcher's famous U-turn in 1982 – it was that this represented over half, indeed nearly 60% of the country's coal industry. The simple reason for this surprise was that British Coal had been following a slow but sure policy of attrition in the industry. Unfortunately the problems with the negotiations over the electricity supply contracts meant that faster action was required because the miners would have discovered that they had lost their jobs from National Power and PowerGen, rather than from British Coal. Whilst one year Britain was thought to be self-reliant on coal the population suddenly found itself facing a much reduced industry because of the death by a thousand cuts that had gone before.

Apart from the deep mines, the prospects for which are covered in more detail in Part III, British Coal also owns and controls the Opencast Executive, although the mining is contracted out to independent companies. Their output has been increasing slowly since it started on an organized basis in 1942 in order to help the war effort. In 1959 a government decision to restrict supplies led to a fall in output, although the 1973/4 oil shock led to a reassessment of this strategy and a target of 15 mt of output by 1985 was set. It was, in fact, attained in 1981 and has remained at or slightly above this level ever since. Most of these operations are totally under the control of British Coal and the company decides how and when mining should take place, often having to plan many years in advance of actual extraction. Indeed, control of a site may last for some 15 to 20 years and in order to maintain output at a high level British Coal needs to purchase between 5000 and 10 000 acres of land annually. The contractors receive a set price for their product which they have to sell to British Coal and they also have to pay a royalty of £4.50/t to £5.00/t based on the amount of coal which is produced. Only about 10% of the open pit mining is carried out by truly private operators which sell their output, largely, to the electricity industry.

Finally, and of relatively little importance, there are the independent underground miners. These small companies are restricted in the amount of coal they can produce because of the statutory limitation on the number of miners they are allowed to employ underground. This goes back to the time of nationalization in 1947 when mines employing less than 30 men underground were excluded from takeover by the government because they were deemed to be too small. In order to prevent the startup of new mines, and the expansion of old ones, the limit on the number of miners who were allowed to work underground in the private sector remained until it was raised to 150 in 1990. This meant that there was no proper competition allowed to British Coal and it is one of the peculiar factors of Mrs Thatcher's administration that this restriction was never lifted (although see the comment on BT above).

The other restriction which had been imposed, admittedly during the pre-war Conservative administration, in order to attempt to encourage the mining companies to merge to form larger and more economic units, was the vesting of the coal in the government rather than in the landowner or allowing a separate ownership of minerals from the freeholder. Up until this legislation was introduced all minerals in the country were owned by the freeholder on whose land they were to be found, unless the metal was either gold or silver in which case ownership was vested automatically in the Crown. The ownership of all of Britain's coal reserves meant that British Coal not only had the right to demand a royalty on all coal which was produced, it also meant that it could control the amount of coal produced by restricting the number of mining licences it issued. This meant that the company was able to restrict its market and could attempt to maintain the price of coal at an artificially high level.

On many occasions both members of the company itself and independent mining companies lobbied the government in order to attempt to change the legislation. They were, however, unsuccessful and the reason for this must be that the government was still looking to achieve as high a price as possible from the sale of British Coal rather than allowing the market to provide a price by showing what competition in underground mining could achieve and thereby providing a benchmark against which British Coal could be judged. As a result the country is likely to be left with a pathetic excuse for a coal industry, set to produce only 30 mt in 1994 compared with the massive 291.6 mt which was produced at the peak level of output in 1913.

Part II

The world coal scene

Chapter

8

Worldwide production and future supply

THE QUEST FOR PROFIT

The UK seems to be one of the few countries in the world which is actually closing down its coal industry with the intention of reducing output. The coal producers in Asia, Australasia, South Africa, Latin America and North America are all opening up new mines and increasing output rather than closing existing operations. The major difference between these countries and the UK is that the mines in all of these countries are generally owned and managed by private companies, whilst the UK mines are owned and managed by the government. It has to be asked whether the government's belief in a free market makes sense when all of the free market companies in other countries are following completely the opposite policy of the British government.

One of the main reasons for this opposing viewpoint must be that these companies know that they can produce coal and sell it at a profit whilst British Coal can only produce coal at a profit when it sells its output at artificially inflated prices to the electricity producers. There are two reasons for this. The most important is that British Coal's mines are underground operations, which require large amounts of capital investment, whereas the new mines being planned and opened up around the world are largely open pit operations. The

UK does have open pit mines but their expansion has been constrained since nationalization by government policy, which has always been strongly in favour of underground mines, and by environmental and heritage objections. This is despite the fact that open pit mining can, actually, enhance the environment over the long term as it can lead to the clean-up of an area spoiled by previous mining operations. This can be seen in the contract let to Wimpey Mining in March 1993 which allows it to work on the site of a former deep mine at Rye Hill, near Durham, and recover some 1.2 mt of coal over a three year period. As some of the coal will come from the re-treatment of 3 mt of waste together with the open pit working of the underground operations (which will improve their safety by restricting access) the quality of the land in the area will be improved. However, the open pit operators have been penalized in the past because British Coal has levied royalties on top of their costs in order to make their total cost of production equate to the selling price of British Coal's output. This has restricted the number of opportunities for independent mining companies to start their own operations and has restricted the amount of price competition suffered by British Coal.

It is because of this that British Coal makes such a strong profit from its Opencast Executive, whilst many of the deep mines still lose money. In a totally free market, which did not have an effective monopoly supplier, the open cast mines would be expanded to the detriment of the underground operations. The depreciation/cash flow problem with British mines has already been covered in Chapter 3. Indeed, the vesting of the country's entire coal reserves in British Coal, at nationalization, also allowed it to restrict the output of open pit coal by only licensing a minimum amount of open pit production, whilst concentrating the maximum amount of output in the deep mined sector. This enabled it to maintain its unproductive ways for much longer than it would have been able to last in a free market and is one of the reasons for the relatively low level of productivity which the company suffered until the change in corporate culture in the 1980s started to have some beneficial effects. In other countries coal is seen, correctly, as a national resource but production will only normally be maintained if it is profitable.

WHAT IS COAL?

Before a full discussion of the world trade in coal takes place it is necessary to define what coal actually is and how this determines where it is consumed. Intrinsic with this is the need to consider how coal formed in the Earth and how this has affected its current composition.

As any fan of the dinosaurs will remember, large tracts of the earth have, at various times, been covered with swampland, often in the deltas of rivers which drained into inland seas in much the same way as the Okavango river drains into a massive swamp and marshland in Botswana today. The main feature of these areas is that they tended to sink relative to the surrounding land and the delta therefore slowly filled up with thicker layers of sediment and organic material. The rate of the sinking played an important part in the formation of the coal measures as too fast a rate would have led to the drowning of the vegetation and allowed the deposition of deeper water sediments such as shales and silts.

If the rate at which the swamp sinks is similar to the rate at which new plants grow then the surface of the swamp will stay relatively stable and more vegetation can be added to that which starts to decay beneath the water surface and form the bed on which the new plants grow. In the major basins of the USA, the Commonwealth of Independent States (CIS) and China this occurred and some very thick seams were formed during the local coal ages. These ages did not take place at the same time in different parts of the world because of the local differences in geological conditions, although the need for long term deposition means that much of the hard coal produced today comes from rocks which were deposited some 240 to 370 million years ago. The age and geographical spread of coal deposits is shown in Table 8.1.

As can be seen in the table there is a wide spread of coal deposits around the world throughout a wide range of geological time. There are few if any deposits older than the 370 million years of the carboniferous period because

Table 8.1 The age and geographical spread of world coal deposits

Age (million years before present) and system	Share of total coal reserves (%)	Area covered, in order of importance
0–65 Caenozoic	28.7	Europe, Australia, New Zealand, North America, South America
65–135 Cretaceous	16.7	North America, South America, Europe, New Zealand
135–200 Jurassic	14.3	Asia, Europe, Australia, North America
200–240 Triassic	0.5	Europe, North America
240–280 Permian	24.3	Africa, Antarctica, Australia, Asia, Europe, North America, South America
280–370 Carboniferous	15.6	Europe, North America, Asia

Source: World Energy Conference, Survey of Energy Resources (1980).

of the lack of spread of vegetation at this time. It is also the case that the older the rock the greater the possibility that geological processes will have affected the coal causing problems of faulting and tilting and alteration associated with the high temperatures and pressures of metamorphism. These temperatures and pressures are required to upgrade the type, or rank, of coal from the basic peaty material which is initially formed, to the high grade and quality anthracites which are used in the metallurgical industry. However, if the coal is subjected to too much heat or pressure it may volatilize with the carbon being absorbed into surrounding rocks or, as occurred with the formation of some of the gasfields in the North Sea, may combine with hydrogen to form natural gas. The rank of a coal is largely dependent upon its age because of the length of time necessary for many of the temperature and pressure processes to take place.

In specific terms the higher the temperature and pressure the greater the proportion of carbon in the coal. This is because of the reduction in the non-carbon constituents of the material which will, effectively, be squeezed out during the metamorphic processes. One of the disadvantages of metamorphism, however, is that it may occur with an introduction of fluids from other rocks. These fluids often contain sulphur and iron and leave deposits of iron pyrites in the coal. This is often a problem and leads to the need for flue gas desulphurization (FGD) equipment in many coal-fired power stations. However, because of the increase in the energy content of the coal which results from the metamorphism, coal consumers may be willing to bear the additional expense of FGD equipment because less coal needs to be burned to produce the same amount of energy. A range of US coals of different age groups is shown in Table 8.2.

The ash content of the coal is also an important consideration and depends upon the specific environment of deposition. If the area was often subject to flooding then there may well have been an influx of sediment which would not have been removed by any of the metamorphic processes briefly

Table 8.2 A range of US coals of different age groups

Age	Moisture (%)	Volatile matter (%)	Fixed carbon (%)	Ash (%)	Sulphur (%)	Energy (MJ/kg)
Carboniferous	10.2–2.4	41.2–5.2	81.3–39.4	12.3–4.2	3.3–0.7	33.56–25.45
Cretaceous	10.7–5.0	42.4–35.8	50.4–38.1	15.0–6.4	1.1–0.6	29.64–23.73
Tertiary	45.9–5.0	35.8–7.2	76.7–22.1	15.7–4.4	2.7–0.3	29.45–13.35

Source: Coal Geology and Coal Technology.

outlined above. In this instance the coal will have a higher ash content and, as the majority of the ash will be inert, a lower energy content than an otherwise similar coal. Additionally, the ash also needs to be disposed of and the increase in the number of environmental restrictions in the USA is starting to become a problem for some of the electricity utilities, particularly those in the north-east of the country.

Most coals, however, are classified according to their carbon content and the specific amount of energy which they contain as a result of this. The greater the amount of carbon the higher the effective rank of the coal and the higher the average energy content of the material. Table 8.3 shows the coal classification of the American Society for the Testing of Materials (ASTM) which is generally accepted as the correct basis for determining the name of a coal, although British Coal has its own nomenclature. It will be noticed that coals of higher ranks are classed in terms of their carbon content whereas coals of lower ranks are classed in terms of their energy content. High carbon coals are generally used in metallurgical processes where the carbon is primarily used as a reducing agent, for instance to take the oxygen out of iron oxide in the production of iron and steel, rather than specifically as a source of

Table 8.3 The coal classification of the American Society for the Testing of Materials

Class	Group	Fixed carbon (%)	Volatile matter (%)	Calorific value (MJ/kg)
Anthracite	Meta-anthracite	98–100	0–2	
	Anthracite	92–98	2–8	
	Semi-anthracite	86–92	8–14	
Bituminous	Low volatile bituminous	78–86	14–22	
	Medium volatile bituminous	69–78	22–31	
	High volatile A bituminous	<69	>31	>32.6
	High volatile B bituminous			30.2–32.6
	High volatile C bituminous			26.7–30.2
Sub-bituminous	Sub-bituminous A			24.4–26.7
	Sub-bituminous B			22.1–24.4
	Sub-bituminous C			19.3–22.1
Lignite	Lignite A			14.7–19.3
	Lignite B			<14.3

Source: Coal Geology and Coal Technology.

heat. However, there are other factors which have to be taken into account when considering whether or not a coal will be suitable for use in the metallurgical industry. These criteria are mainly related to the coal's 'caking' power which is its ability to cake or fuse when heated. As only some of the organic constituents of the coal contribute to this caking process coals of a similar carbon and ash content may fall into different categories. Additionally coals of too high a rank will not have a sufficiently high level of organic constituents and will therefore not be suitable for use in the metallurgical industries.

WORLD PRODUCERS

In 1991 Britain was the tenth largest coal producer in the world. In 1994, when British Coal's output will have fallen by some 40 mt as a result of the pit closures, the country will have fallen to about thirteenth place, assuming that there is no increase in production elsewhere. However, as will be seen in the following section there is likely to be a continued increase in the amount of coal produced around the world between now and the end of the century.

By far the largest coal producer in the world, some 200 mt greater than its nearest rival, is China. The USA stands in second place, followed by the CIS in third. Of these states Russia and the Ukraine make up the largest part with mines in two vast coal basins in the two countries. The top 20 world coal producers in 1991 are shown in Table 8.4, together with their 1991 output and the distribution of output between hard coal and lignite or brown coal.

The ranking of world coal exporters is vastly different from that of world producers because of the large amount of coal which is consumed in the home market. It is also interesting to note various disparities between the geographical areas of exports and consumption. For instance, whilst China is the world's largest producer of coal and is situated so close to Japan, which is one of the world's largest consumers, only 1.8% of China's entire production is destined for the export market. Australia stands as the world's largest single exporter of coal; at 115.5 mt this represents 53.5% of the country's total production of both hard coal and brown coal or lignite. The export market for lignite is not well developed as the relatively low energy content and hence price of the commodity means that it is normally cheaper to utilize imported hard coal or home supplies of lignite than to import lignite from overseas. Assuming that Australia's exports are entirely made up of hard coal the proportion of exports to production rises to a massive 72.2%. The world's

Table 8.4 The top 20 world coal producers and their 1991 output of hard coal and lignite

Country	Total (mt)	Share of world (%)	Hard (mt)	Lignite (mt)
China	1090.0	24.4	976.0	114.0
USA	902.0	20.2	609.0	293.0
CIS	569.0	12.7	417.0	152.0
Germany	345.5	7.7	66.1	279.4
India	223.0	5.0	211.0	12.0
Australia	215.6	4.8	166.3	49.4
Poland	209.6	4.7	140.3	69.4
South Africa	177.5	4.0	177.5	0.0
Czechoslovakia	98.8	2.2	19.5	79.3
Great Britain	96.2	2.2	96.2	0.0
Canada	71.1	1.6	61.3	9.8
Yugoslavia	71.0	1.6	0.0	71.0
Greece	50.6	1.1	0.0	50.6
Turkey	48.2	1.1	4.6	43.6
Korea DPR	43.3	0.9	43.3	0.0
Spain	36.0	0.8	14.9	21.1
Romania	32.4	0.7	3.5	28.9
Bulgaria	29.1	0.7	0.1	29.0
Colombia	23.0	0.5	23.0	0.0
Hungary	17.0	0.4	1.7	15.3
Rest of world	123.5	2.7	95.4	28.2
Total world	4472.3	100.0	3126.5	1345.8

Source: Mining Magazine.

largest coal exporters and the share of hard coal production which was exported in 1991 are shown in Table 8.5.

The self-sufficiency of the former communist bloc is clearly evident from these statistics. Nevertheless, there is a slow change which is becoming evident as the countries reduce their previously self-imposed barriers to world trade. This means that the world can expect the former Soviet Union countries, in particular, to increase the share of coal production that is exported in order to satisfy their growing need for foreign exchange. This is necessary in order to prevent the creation of a massive trade deficit with the West and has been evident in these countries' exports of both gold and nickel in recent years. It has undoubtedly had a major influence on the recent decimation of the prices of both of these metals on the international markets. Although the CIS mines

Table 8.5 The big coal exporters' share of hard coal production

Country	Coal exports, 1991 (mt)	% of hard production
Australia	120.2	72.2
USA	98.8	16.2
South Africa	47.8	26.9
Canada	34.1	55.6
CIS	28.1	6.7
China	18.8	1.9
Poland	18.0	12.8
Colombia	14.7	63.9
Indonesia	6.5	45.5
Germany	3.3	5.0
Venezuela	2.1	84.0
Great Britain	1.8	1.9
Rest of world	6.7	N/A
Total world	400.9	N/A

Source: Mining Magazine, OECD.

are highly dangerous and uneconomic and the CIS is in need of as much of its own indigenous energy as it can produce, it is likely that the level of exports from these countries will increase in the future. The introduction of Western oil companies with modern production technology has been a recent development and it must only be a matter of time before the same happens to the countries' coal industry.

This, however, may not lead to an increase in the amount of coal which is produced as the new equipment may only increase productivity and safety rather than have any effect on production (i.e. it will lead to a reduction in the numbers of people required to produce the same amount of coal). Indeed, as can be seen from the experience of the UK coal mining industry increased mechanization increases the capital intensity of a mining operation by increasing fixed costs and leading to a reduction in the variable cost of labour in an unmechanized operation. A similar situation exists in China where the coal mines employ vast numbers of the population and yet are highly dangerous places to work. Recent production of coal in the CIS is shown in Figure 8.1, although the reliability of some of the figures must be in question as some are US Central Intelligence Agency (CIA) estimates which, as far as gold production is concerned, have been shown to be very inaccurate.

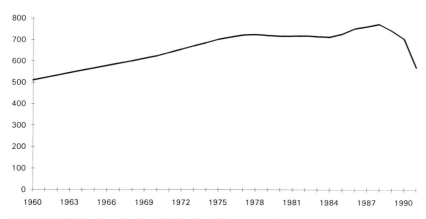

8.1 USSR/CIS coal production 1960–91 (million tonnes)
Sources: Novosti Press Agency, Mining Annual Review.

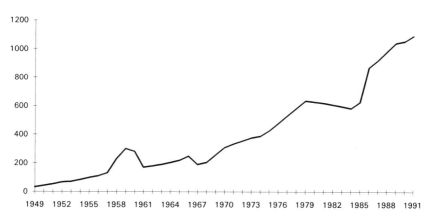

8.2 Chinese coal production 1949–91 (million tonnes)
Sources: China's Economy, Mining Annual Review.

A certain amount of caution must also be utilized when collating Chinese production statistics. Only 'official' statistics will ever be released by the Chinese government and may bear little relation to the actual amount of production from the mines within the country. Nevertheless, the country has huge reserves of the commodity and output is broadly in balance with demand, although a lack of sufficient energy supplies has been one of the reasons why the country's industrial growth rate has slowed in the past. Recent Chinese coal production is shown in Figure 8.2.

India is also a country with a vast coal mining industry which is largely state controlled. In this country too there is very little coal sent for export with the majority of the country's output being consumed in the home market. India also employs vast numbers in its mines and a recent count showed that the industry was paying miners who did not even exist, or who had either died or retired many years previously!

FUTURE SUPPLY

Coal will always be produced around the world and, with an expanding world economy, it is likely that coal will continue to be important despite the environmental considerations which will weigh against it. This is particularly because it is a relatively cheap source of energy which does not need to be treated or refined before it is used. The generation of heat both for domestic use and for electricity generation and the consumption of coal in the metallurgical industries are likely to remain the most important areas of consumption and the plentiful supply of coal means that it is unlikely to be usurped by any of its competitors. As such it is possible to assume that as much coal will be consumed as can be produced, although this may have short term adverse implications on the price of the commodity (as is currently the case).

As mentioned at the start of this chapter most of the world's coal producers are intending to, or have recently completed, expansions in their productive capacity and the amount of coal which will be produced in the future can be expected to increase. In particular China was intending to increase its coal production marginally in 1992 to some 1100 mt from the 1090 mt produced in 1991. This is only the start of a proposed expansion programme which, it is hoped, could see China's production of coal doubling over this decade. At present China's coal consumption is estimated to generate some 10% of the world's entire carbon dioxide output and a doubling of coal consumption would have a devastating effect on the world environment. Nevertheless, much of this may be scare tactics as China's planning and the actual outturn of events are not necessarily the same, as shown by the results of many of the five year plans the Chinese government has introduced in the past.

Elsewhere in the world the South African industry is poised for a major coal expansion as it returns to the international market without the stigma of apartheid. This is particularly important to the South African producers because they were previously forced to sell their output at a discount to the ruling international price. The country is, however, constrained by the size of its major coal port at Richards Bay. This port has the capacity to export only some 46 mt of coal annually and there are difficulties about the cost of, and who should pay for, a proposed expansion to a capacity of around 53 mt/year. The coal producers are all joint venture partners in the coal terminal but not all of them want to pay for the expansion. This is because some of them would not be able to increase output to the level necessary for them to make use of the additional capacity they would be allotted. Nevertheless, an expansion of the terminal is to be expected at some stage over the next few years which will lead to an increase in South African exports.

Despite all of this action occurring elsewhere in the world it is likely that the Latin American countries of Colombia and Venezuela will become the major players in the international coal market of the next few years. This is because of the vast reserves which are situated in these countries and the relatively efficient infrastructure and low labour costs which will enable the coal to be exported cheaply. In particular Venezuela, with its Gulf of Mexico coastline, is well situated to take advantage of the European market and has a massive expansion of its coal mining industry planned over the next few years. As the development of South American economies progresses it may be that some of the coal is exported to neighbouring countries although much of it will be sold to European consumers.

One factor which may influence the decision of neighbouring countries to import coal to meet their growing energy demands is the destruction of the Amazon rainforests and the denudation of the landscape which leads to a further reduction in rainfall. If the countries suffer from an energy shortage, which sometimes seems to have been the case in recent years, then it is possible that the destruction of the forests for firewood may be slowed down and the consumption of coal may increase. This was the first reason for the growth in consumption of coal during the eighteenth century in England, especially around London where the destruction of forests meant that there was a marked shortage of firewood for heating and cooking.

In addition to Colombia and Venezuela Indonesia is attempting to break into the international coal market. In Indonesia CRA, the 49.9% owned associate of RTZ Corporation, and BP have recently opened the Kaltim Prima coal mine under a 50:50 joint venture. The mine should produce at a rate of some 7 mt/year and has already affected the price of coal in the Asian market.

Other countries are also expected to increase their energy requirements and this is likely to lead to an increase in the demand for coal in the future. The USA is, perhaps, the only country where there may be a fall in demand (apart from the UK) as the increasingly stringent environmental legislation leads to greater restrictions on the sulphur content of the coal which can be burnt in the country. The highest sulphur coals in the USA come from the

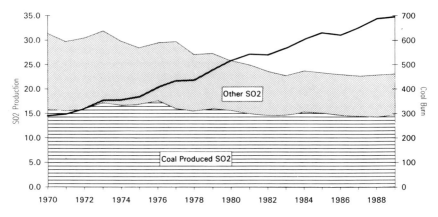

8.3 US SO$_2$ production and utility coal burn 1970–89 (million tonnes)

north-eastern end of the Appalachian belt which stretches from Pennsylvania in the north-east to Alabama in the south-west. The sulphur content of these coals can approach 2.3% as at the Cyprus Minerals Emerald Mine; the coal produced by Cyprus Minerals has to be blended with low sulphur coal from other mines in order to stay within the country's emission control legislation. The emissions of sulphur dioxide in the USA and the share of those emissions produced by coal-fired electricity generators is shown in Figure 8.3, together with the coal consumption of the electricity generators. This shows clearly that despite an increase in the amount of coal burnt, the country has been able to reduce emissions considerably since the peak in 1976.

Finally, there are small developing countries such as Botswana which have a slowly emergent economy. Botswana's hard coal production in 1991 was only 790 000 t, and there is no lignite production in the country. All of this production came from one mine which has a capacity to produce at a rate of around 900 000 t/year. However, Botswana has reserves of a massive 12 billion tonnes situated in two separate coalfields, the Morupule Field with reserves of 6–8 billion tonnes and the Mamabula Field with reserves of around 4 billion tonnes. In the future the country, which remains heavily dependent upon diamonds and De Beers' efforts to maintain stability in the diamond market, hopes to increase its coal production and export a less expensive form of carbon.

Projections of coal production until 2010 are shown in Table 8.6.

Table 8.6 Projections of coal production until 2010

Country	1991	1992	1993	1994	1995	2000	2010
	million tonnes						
China	1090.0	1100	1130	1160	1200	1300	1500
USA	902.0	900	910	920	940	1000	1100
CIS	569.0	500	450	450	450	600	700
Germany	345.5	300	300	280	280	250	200
India	223.0	225	235	245	250	300	400
Australia	215.6	220	225	235	250	300	350
Poland	209.6	200	200	210	210	220	250
South Africa	177.5	180	182	185	190	200	225
Czechoslovakia	98.8	100	100	100	100	100	100
Great Britain	96.2	70	60	60	60	60	60
Canada	71.1	65	68	70	70	75	80
Yugoslavia	71.0	30	30	40	50	70	70
Greece	50.6	52	52	52	52	57	60
Turkey	48.2	50	55	65	80	100	120
Korea DPR	43.3	43	45	48	50	60	70
Spain	36.0	35	34	34	34	30	25
Romania	32.4	32	32	32	35	38	40
Bulgaria	29.1	30	30	30	30	34	38
Colombia	23.0	25	30	30	40	50	60
Hungary	17.0	17	17	18	18	20	20
Korea ROK	15.1	16	16	16	18	20	30
Thailand	14.7	15	15	15	16	18	25
Indonesia	14.5	15	20	22	25	35	50
Venezuela	2.5	4	5	8	10	25	50
Rest of world	76.7	80	82	85	90	100	120
Total world	4472.3	4304	4323	4410	4548	5062	5743

Chapter

9

Worldwide demand

WHO USES COAL?

Almost every country in the world which can afford it uses coal as a source of energy. This can be in the generation of electricity, which is then sent down power lines and more effectively distributed to the population, in larger metallurgical works or in private consumption. The major limiting factor in most of the world is the source of the coal and the resultant cost of its supply to the consumer. This has also to be compared with the relative cost of other sources of power as the debate over gas utilization in Britain at present shows. In the USA, in particular, and spreading slowly across Europe, the environmental cost of using coal as a source of energy is also gaining recognition. This cost is said to be significant because coal generates both carbon dioxide, a suspected greenhouse gas, and sulphur dioxide which is the prime source of acid rain.

As Figure 9.1 demonstrates, the use of coal in many applications in the UK, which is an indication of its role elsewhere in the world, has declined markedly since the end of World War II. The main reason for this is the spreading use of other sources of power, most notably the replacement of coal-fired steam engines with first oil-fired steam engines and later diesel electric and simple electric engines and the spread of the internal combustion engine

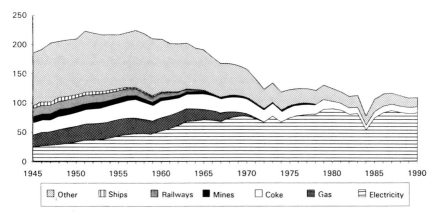

9.1 Consumption of coal in the UK by end use 1945–90 (million tonnes)

generally. The decline in the steel industry has also led to a fall in the demand for coal in the UK despite British Steel's position as one of the lowest cost producers in Europe. Other steel producers, notably those in Spain and Germany, are able to continue only by virtue of the massive state subsidies they receive. It should further be noted that both countries heavily subsidize their indigenous coal industries. Despite this the use of more efficient techniques means that less coal is now used. Overall, therefore, coal usage has come to rely on the electricity generation industries and this is one reason why some coal producers tie up their production over long term contracts with electricity utilities, particularly in the USA.

The increasing acceptance of long term supply contracts with the electricity generators is a direct result of the oil crises of 1973/4 and 1979. The reasoning was simply that the generators wanted to ensure that they did not end up having to pay premium prices as a result of the increased demand for coal from other consumers who were switching away from oil because of its rising price. All forms of energy were rapidly increasing in price at this time and the recent experience of the uranium market is very much a mirror image of the UK coal market today.

Most of the nuclear power stations built around the world in the third quarter of this century (Britain's first station, Calder Hall, opened in 1956) needed a guaranteed source of uranium in order to ensure that they would be able to meet their contractual commitments to supply electricity to their customers. At the time the demand for electricity was forecast to continue to grow dramatically and the forecasts of future uranium consumption now defy belief. The power generators therefore guaranteed to purchase a fixed amount of uranium from the producers at prices linked to an agreed inflation indicator over long term contracts of up to 15 years' duration.

What the forecasters did not foresee, however, was that the oil price rises would lead to slowdowns in growth, and that the amount of energy consumed would not meet expectations, despite the consequent fall in the price of oil. As a result the demand for uranium never met expectations and the squeeze on supply was only shortlived as the contracts were signed up. Since then the price of the metal has plummeted to around US$6-9/lb, whereas some electricity generating companies are still being forced to pay prices of US$60/lb and over for their material on the basis of the contracts they entered into so long ago. As these contracts expire, the uranium producers are being forced to close down their mining operations because they are not profitable at the ruling market price. There are no uranium mines in the UK and so there has been no public outcry at their closure.

At the same time that the uranium consumers in the electricity generation industry were fixing up long term contracts with the uranium producers, the coal producers were signing long term coal supply contracts on similar bases. This also led the coal producers who were lucky enough to enter into these contracts into a period of high and increasing revenues despite the coal price falling in the free market. The expectation of just such a situation led Sir Ian MacGregor, when President of the US mining company Amax, to invest heavily in the US coal industry. Amax has been broadly successful in US coal mining with large cheap mines in the Powder River Basin.

Unfortunately not all US coal producers have been so fortunate, especially as far as long term contracts are concerned. One such company was Cyprus Minerals which was originally spun off as a free distribution of shares from the US oil company Amoco in 1986. At the time of the spin-off many oil companies were getting out of the minerals businesses which they had purchased at the top of the 1979/80 oil price bonanza. Included in the deal, therefore, were all of Amoco's former mining interests, including its non-oil energy interests in the form of its coal deposits. Some of this production had the benefit of being protected from the decline in the coal price by virtue of long term supply contracts with coal consumers and the most notable of these was the contract between Cyprus and LTV, the big US steel and armaments producer. As a direct result of being forced to pay high prices for its coal LTV's energy costs were too high, it was not able to match the low prices of its competitors and it was forced into Chapter 11 bankruptcy.

Such an example is behind Michael Heseltine's assertion that high priced electricity generated from high priced coal will not be good for Britain's economic future. It will mean that the country will not be able to compete effectively in the free market unless there is a continual devaluation of the currency which makes it possible to produce a good and then sell it at a higher sterling price but at a lower US dollar or Deutschmark price. The difficulty further arises because of the need to ensure a level playing field as far as electricity supply is concerned and the huge levy which is forced on the country by the need to decommission the nuclear power stations is a severe handicap to the rest of the industry. Whilst Arthur Scargill may complain that the pit closure programme represents 'the economics of the madhouse' the nuclear industry subsidy is really where the economic madness lies. This is because of the high costs which it will continue to force on the electricity consumer over the foreseeable future. It will therefore distort the market and will be an important factor in reducing the international competitiveness of British industry.

WHERE THEY USE IT

The use of coal is generally restricted by economic considerations, mainly linked to those mentioned at the start of this chapter. The main producers of coal are also the main consumers of it because of the costs of transporting what is really a very bulky and immovable source of energy. It does not have the advantage of oil or gas which can be piped and pumped around the country or the world. Nor does it have the advantage of gas which can be liquefied into a smaller vessel and shipped around the world. Uranium, also, is a relatively easy fuel to transport in terms of its bulk, although the radiation risks mean that safety and security costs also need to be taken into consideration. Electricity is also a very easy form of energy to transmit, although the further the distance the greater the loss of power because of the inherent resistivity of the steel, aluminium or copper cables used in the grid. Hydro-electricity is dependent, largely, on countries with mountainous topography and would not be an efficient source of energy in the southern part of Britain.

Generally, however, any country will adopt a varied energy policy, if at all possible, because of the need not to become too dependent upon one source of power. In such a case the country could become 'hostage to energy ransom' as Neil Kinnock, the former leader of the Labour Party, put it in the questions to Michael Heseltine on 13 October 1992. This, normally, is even though the chosen form of energy is more expensive than an imported alternative, and this extra cost is accepted because of the need to ensure a security of supply. This therefore results in the continuing use, or the commission of, new sources of energy despite the huge costs they will force a country to bear. In France a massive proportion of the country's electricity is generated in nuclear power stations and the high cost of this source of electricity has to be borne by the population. The reason for the country's insistence on nuclear generation, as opposed to all other sources of electricity, is that it has no other indigenous sources of energy apart from some small coal mines in the north-east (which have now largely been closed down) and small deposits of oil and gas in the Paris basin. Indeed, France became an importer of British coal in the reign of Edward III when he authorized exports of the fuel. Other imported means of electricity generation are used by France but the country prefers to be as self-dependent as possible, as can be seen by the French government's insistence (whatever its political persuasion) that the nuclear industry is maintained. Cogema, the state company which runs the nuclear industry, also acquires significant equity stakes in existing or potential uranium producers, wherever in the world they are based, in order to attempt further to guarantee supplies.

Because of these restrictions and requirements any producer and supplier of coal has to meet exacting terms if a consumer will commit to purchase coal over a long period of time whether for electricity or for steel production. One of the main problems covers the security of supply and this is one of the reasons why a large amount of South African coal still has to be sold on the Amsterdam/Rotterdam/Antwerp spot market rather than being sold on longer term contracts directly to consumers. In the past, with the political stigma imposed on South African output, it probably suited the country to avoid direct marketing and to sell through intermediaries so that the end buyer did not necessarily know where the coal originated. Now, following the dismantling of apartheid, more longer term contracts are likely to be entered into although the worry over the future political complexion of the country and the concern that nationalization may lead to disruptions to supply are likely to deter some consumers from relying on the country as a prime source of energy supply for some time to come.

One of the strange facts which arises as a result of the problems over the source of coal and where it is to be used occurs in China. As stated in the previous chapter, the country is, by a large margin, the world's largest producer of coal. This, however, has not stopped it from importing coal in the past because of the problems of distributing the coal from the mines which tend to be in the north of the country to some of the consuming areas in the south. It is sometimes cheaper for the Chinese to import Australian coal than it is for them to rely on the country's reportedly poor infrastructure to move it from one place to another.

The other problem which arises over the source and use of coal as opposed to any other form of energy is the use to which it is to be put. Crude oil, whilst varying to some degree in texture and composition, must still be cracked to produce the components which are used in its various applications. Coal, on the other hand, varies considerably from location to location and different types or ranks of coal can only be used in certain applications. It is also not economically possible to break coal down into its constituent parts to enable it to be used in any application whatever its original composition (although the South African company Sasol would have us think otherwise, it used to receive a massive grant from the South African government to continue its work into the oil from coal technology which it has pioneered, and had little problem meeting the initial capital costs of the plant and equipment it uses).

This is the prime reason why the UK imports German coal which is twice as costly to produce as British coal. The German imports are used in specific applications which cannot be supplied by coal mined from British pits. This may soon be true of another industry in the UK, Coalite, owned by Anglo United, which is dependent on specific coal from the adjacent Grimethorpe colliery to enable it to make the end product with which the company's Coalite subsidiary is synonymous.

THE INTERNATIONAL COAL MARKET

The international market in coal is therefore restricted to a few key players in a relatively small number of countries. All other producers either use all of their own production locally and are unable to export, or the cost of production is so high that their exports are not economic as cheaper coal can be purchased elsewhere. As mentioned in Chapter 8, the world exporters of coal are relatively few and limited, although the majority are currently in Organization for Economic Co-operation and Development (OECD) countries, particularly Australia and the USA. Of the non-OECD countries South Africa is the largest single exporter and accounted for some 12% of the total world market in hard coal of 400.9 mt in 1991. This market is split about 45%:55% between coking coal and steaming coal which are used in the metallurgical and power generation industries respectively.

The greater demand for steaming coal is related to the huge energy demands of the European OECD countries which accounted for half of the steam coal demand in 1990 and 1991. The second largest consumers of steaming coal are the developing Asian economies, although they accounted for only some 18% of international steaming coal trade. Conversely the

Japanese are the world's largest consumers of coking coal, accounting for 38% and 40% of world coking coal demand in 1990 and 1991 respectively. The international trade in coking coal is larger than its relative share of production because of its relative scarcity and importance in the world's metallurgical industries.

This demand is fuelled by the large Japanese metallurgical industries which need to import much of the coal they consume. The Japanese were once large coal producers in their own right but from a total workforce of 348 000 people the Japanese industry now only employs some 5000 people in five collieries. This has led to a marked need for imports of coal into the country to feed the metals industry, although the country's imposition of import tariffs on finished metals means that it is still cheaper for the Japanese to go to this extra expense than it is to import the finished metal for use in its industrial base.

European OECD countries make up the second largest importers of coking coal in order to fuel their own metallurgical industries. Most European imports originate from the USA, as opposed to the Australian dominance of the Japanese market. This clearly shows the cost benefit of being near to the source of the material as the greater the distance the greater the costs of transport. In the case of Australian exports to Japan much of the coal is sold free on board (FOB) the local Australian port providing the Japanese with the ability to react to their own transportation costs. This is not the preferred method of sale for the Australians who would prefer to sell the coal after taking account of carriage insurance and freight (CIF) as this provides them with a potential additional profit margin. It is also the case that the Japanese are very much aware of the international coal market and that most of the trade is carried out in US dollars. The Australians are therefore highly exposed to the local fluctuations in the Australian currency against the dollar and the Japanese take full advantage of this when conducting their annual contracts. This means that a rise in the Australian dollar would lead to a fall in any Australian dollar contract price negotiated between the two parties, assuming that the international price of coal, in US dollars, remained stable.

The major exporters and importers of both coking and steaming coal in 1990 and 1991 are shown in Tables 9.1–9.4.

FUTURE DEMAND TRENDS

The future of coal demand is largely restricted to the advent of a new cycle of economic growth in the industrialized economies and

Table 9.1 Major exporters and importers of coking coal in 1990

	North America	OECD Europe	Japan	South America	Asia	Africa & Middle East	Former CPEs*	Balancing item	World total
Australia	—	10.0	29.6	1.1	15.2	0.3	1.6	−0.9	56.9
Canada	—	2.8	17.6	1.1	5.9	—	—	−0.5	26.9
USA	5.0	28.4	9.6	5.9	3.3	1.1	2.6	1.7	57.6
Others	—	3.3	0.2	—	0.1	—	0.1	0.7	4.4
Total OECD	5.0	44.5	57.1	8.1	24.4	1.4	4.3	0.9	145.7
Poland	—	2.6	—	2.4	0.3	0.1	0.9	—	6.3
CIS	—	1.8	5.5	—	1.5	0.2	8.2	—	17.2
China	—	—	1.3	0.1	1.9	—	0.4	—	3.7
Colombia	—	—	0.1	—	—	—	—	—	0.1
S. Africa	—	0.2	3.4	—	—	—	—	—	3.6
Others	—	0.9	0.2	—	—	—	—	—	1.1
Total non-OECD	—	5.5	10.5	2.5	3.7	0.3	9.5	—	32.0
TOTAL	5.0	50.0	67.6	10.6	28.1	1.7	13.8	0.9	177.7

* Centrally planned economies.
Source: OECD.

119

Table 9.2 Major exporters and importers of coking coal in 1991

	North America	OECD Europe	Japan	South America	Asia	Africa & Middle East	Former CPEs	Balancing item	World total
Australia	—	12.2	35.7	2.5	17.0	1.5	0.7	-3.9	65.7
Canada	—	2.9	17.5	1.6	6.5	0.3	—	—	28.8
USA	4.9	28.0	10.0	6.8	3.1	1.7	2.7	1.4	58.6
Others	—	2.2	0.4	—	0.2	—	—	—	2.8
Total OECD	4.9	45.3	63.6	10.8	26.8	3.4	3.5	-2.3	156.0
Poland	—	2.3	—	1.5	0.3	0.1	2.1	—	6.3
CIS	—	0.7	3.7	—	0.7	0.1	4.4	—	9.6
China	—	—	1.7	0.1	2.0	—	0.4	—	4.2
Colombia	—	—	0.3	—	—	—	—	—	0.3
S. Africa	—	0.1	3.3	—	—	—	—	—	3.4
Others	—	0.9	1.8	—	—	—	—	—	2.7
Total non-OECD	—	4.0	10.8	1.6	3.0	0.2	6.9	—	26.5
TOTAL	4.9	49.3	74.5	12.4	29.8	3.6	10.4	-2.4	182.5

Source: OECD.

Table 9.3 Major exporters and importers of steaming coal in 1990

	North America	OECD Europe	Japan	South America	Asia	Africa & Middle East	Former CPEs	Balancing item	World total
Australia	—	9.3	24.5	0.1	12.8	0.5	—	2.0	49.2
Canada	0.9	0.8	1.3	0.1	1.3	—	—	-0.3	4.1
USA	9.1	20.3	1.3	0.7	4.7	1.2	—	1.1	38.4
Others	—	8.0	—	—	—	0.4	0.3	-1.1	7.6
Total OECD	10.0	38.5	27.2	0.9	18.8	2.1	0.4	1.4	99.3
Poland	—	10.8	—	—	—	—	10.1	0.9	21.8
CIS	—	7.3	2.8	—	0.1	—	12.1	—	22.3
China	—	2.8	3.3	—	7.1	—	—	—	13.2
Colombia	1.3	8.3	—	3.0	0.6	0.6	—	—	13.8
S. Africa	—	26.0	1.4	1.3	15.0	2.6	—	—	46.3
Others	0.3	2.8	1.3	—	—	—	—	—	4.4
Total non-OECD	1.6	58.0	8.8	4.3	22.8	3.2	22.2	0.9	121.8
TOTAL	11.6	96.5	36.0	5.2	41.6	5.3	22.6	2.3	221.1

Source: OECD.

Table 9.4 Major exporters and importers of steaming coal in 1991

	North America	OECD Europe	Japan	South America	Asia	Africa & Middle East	Former CPEs	Balancing item	World total
Australia	—	10.7	23.9	0.3	13.9	0.5	0.1	5.1	54.5
Canada	0.8	1.1	1.3	0.1	1.8	—	—	0.2	5.3
USA	7.5	27.9	2.0	0.4	5.0	1.2	0.2	-4.0	40.2
Others	—	6.0	—	—	—	0.1	—	1.0	7.1
Total OECD	8.4	45.7	27.3	0.8	20.6	1.9	0.3	2.1	107.1
Poland	—	9.7	—	—	—	—	2.0	—	11.7
CIS	—	8.1	2.6	—	—	—	7.8	—	18.5
China	—	3.2	3.9	—	7.5	—	—	—	14.6
Colombia	1.7	11.6	0.1	0.3	—	0.7	—	—	14.4
S. Africa	—	26.8	1.9	1.3	11.8	2.6	—	—	44.4
Others	0.5	5.5	1.7	—	—	—	—	—	7.7
Total non-OECD	2.2	64.9	10.2	1.6	19.3	3.3	9.8	—	111.3
TOTAL	10.6	110.6	37.5	2.4	39.9	5.2	10.1	2.1	218.4

Source: OECD.

the continuing growth of the developing nations. Against these factors will have to be set the growing level of environmental awareness which will tend to reduce fossil fuel demand through the introduction of carbon taxes and will attempt to make industry more efficient in its use of energy. Therefore it is to be expected that world energy consumption will grow at a slower rate than the overall rate of growth of the combined world economy.

Within this total energy consumption picture the use of coal has to be considered on its individual merits. It is currently less energy efficient to burn coal in power stations than it is to burn gas in the newly developed combined cycle gas turbines (CCGT), especially where new CCGT plants are being constructed in combined heat and power (CHP) developments. In the short term, therefore, it is unlikely that there will be a marked upturn in coal demand at the expense of any other fuel medium and it is to be anticipated that coal consumption will, in fact, show a relative, although not necessarily an absolute, fall in demand. This is likely to be particularly the case in the industrialized world which is more able to afford the greater cost of environmental awareness.

In the developing countries the use of coal can be expected to grow at a rate similar to the growth of the local economy. This is particularly because many of these countries have indigenous coal deposits and, therefore, do not have to spend large amounts of precious foreign exchange to import oil. As a result they may show a greater use of coal in the attempt to increase their economic growth, although this is not in the best interests of the environment. The greater use of local coal supplies may also become prevalent in the coming years because of the lack of precise knowledge about where a carbon tax will be imposed.

Indeed, all of the OPEC countries could add the US$10/barrel carbon fuel tax being debated by the European Community overnight if this was required (as they have shown in the past). This, of course, would not be politically acceptable to the oil importing nations who would not wish to provide OPEC with a bonus. Apart from this the addition of US$10 to a barrel of oil has been calculated to increase the cost of petrol at the pump by only some 10% because of the high refining and marketing costs associated with petrol. Such a small rise would not have much of a deterrent effect on the use of oil and a consumption or purchase tax on the end user rather than a purchase tax on the supplier/refiner is much more likely to be imposed.

Such a tax would also have ramifications in the European electricity market and implications for the coal industry. If a carbon tax is introduced it would have to be imposed on the burning of coal in power stations in much the same way as an increase in the tax on petrol at the pump. As a result it would have to be a severe tax if it is to deter consumers from using electricity. Apart from this the day of the intelligent electricity meter is many years away and consumers are not yet able to decide whether they wish to purchase highly taxed coal- or oil-produced electricity, untaxed wind- or hydro-produced electricity, or whether they want to suffer any of the political fall-out from using nuclear-generated electricity.

There are therefore many question marks hanging over the future use of coal in the world, although the international trade represents only a relatively small part of the total amount of coal burned. Nevertheless, it is expected that the international trade in coal will continue to grow, albeit at a slower rate than the world economy, until the end of the century. Beyond this the use of coal will depend on the imposition of carbon and other taxes, and on the anticipated increase in the prices of both oil and gas which may tend to make coal a more economical fuel despite the high tax burden which it is likely to suffer.

UK EXPOSURE TO THE INTERNATIONAL COAL MARKET

The United Kingdom both imports and exports hard coal and therefore has some exposure to the international market. Most, however, of the country's exposure is to speciality coals which are either peculiar to this country and needed overseas or are coals which are not produced locally but are required by British industry. This is, specifically, one of the conundrums of the closure programme proposed in October 1992 as one mine in particular, at Betws Drift in South Wales, is a speciality anthracite producer which does not have any other competition within this country, although there are other producers of similar anthracite overseas.

In 1983 British exports of coal were higher than the country's imports. This was the last year that the UK had a net trade credit in coal as the debilitating effects of the 1984/5 miners' strike led to the final death knell in the majority of the country's export markets. It forced the UK's customers to look elsewhere for their supplies and the country was unable to regain these markets following the eventual end of the strike. The strike also led to a marked increase in British coal imports, which doubled from their 1983 level in 1984 and had trebled by 1985. The country imported nearly 4.5 times the 1983 level in 1991 and the trend in British imports of coal is likely to continue to increase in the future as the intention of British Coal is to reduce its output because of the company's belief in the uncompetitive nature of British production (see Part III).

As far as the British distribution of imports is concerned, it was coking al imports which were specifically in demand during the majority of the 80s. It was only in 1991 that there was a massive increase in the demand r steaming coal, which increased by some 68% over the level for 1990 as the pending nearness of the renegotiation of new contracts with the power nerators and the desire to run down coal stocks at the mines led to a further cline in British Coal's output. That British Coal's stocks are reported to be large and yet there was such a large increase in the level of imports begs e question why British Coal did not sell at least some of its stocks at the arket price rather than at its contracted price.

The other peculiarity concerning the imports of steaming coal is the 7 000 t imported from Germany. In the Parliamentary debate following his atement on the proposed closures Michael Heseltine, the President of the ard of Trade, stated that Britain 'imported no German coal for electricity neration' and implied that this steaming coal is used for other purposes. It ems, however, that there cannot be a substitute produced in this country for rman coal which is very heavily subsidized by the German government. The el of subsidies has been variously estimated at between £1 billion and £4.5 lion depending upon which specific calculations are included and has been id to equate to a German cost of production of some US$285/t compared h the current British cost of some US$65/t. This seems to be one of the quities resulting from the lack of a pan-European energy strategy despite first soundings of the European Community coming from the formation of European Coal and Steel Commission. German coal industry subsidies are own in Table 9.5.

The levels of hard coal exports and imports of both coking and steaming al between 1983 and 1991 are shown in Tables 9.6–9.8. Of particular note the tables are the dramatic falls in the level of exports as a result of the 84/5 miners' strike; in many cases, the markets were either lost completely vitzerland) or never fully recovered (Germany and France). As far as the

Table 9.5 German coal industry subsidies, 1980–92

Year	1980	1986	1990	1992
Subsidy (DM m)	4470	4563	8599	8401
£ equiv. @ 2.40 DM: £1	1863	1901	3583	3500

Source: Financial Times.

125

Table 9.6 Exports of UK hard coal, 1983–91

	1983	1984	1985	1986	1987	1988	1989	1990	1991
Belgium	105	73	44	51	7	13	45	103	106
Denmark	1571	612	1278	1024	1023	426	521	366	175
Finland	765	114	2	46	64	64	186	204	72
France	1566	778	231	425	217	256	210	271	346
Germany	624	122	191	200	211	105	93	214	244
Greece	3	25	13	0	0	0	0	0	0
Iceland	15	8	19	6	0	13	19	18	1
Ireland	428	309	392	482	331	359	331	247	268
Netherlands	318	71	14	38	2	8	10	25	11
Norway	68	24	12	74	111	107	128	145	73
Portugal	134	5	111	199	225	170	153	190	118
Spain	11	21	3	3	0	0	33	138	256
Sweden	201	34	65	89	123	50	38	87	29
Switzerland	169	83	0	0	0	0	0	0	0
Other OECD	6	6	1	0	1	3	0	1	0
Total OECD	5984	2285	2376	2637	2315	1574	1767	2009	1699
Brazil	0	0	0	0	0	0	6	5	1
Asia/Oceania	0	0	0	4	3	2	1	1	3
E. Europe/CIS	0	0	1	2	0	3	0	0	1
Israel	0	0	0	54	0	0	0	0	0
Egypt	0	0	0	0	0	0	1	1	0
Africa/Middle East	351	150	178	52	23	144	285	282	105
Others	5	5	3	0	0	14	12	9	3
Total non-OECD	356	155	182	112	26	163	305	298	113
TOTAL	6340	2440	2558	2749	2341	1737	2072	2307	1812

Source: OECD.

import statistics are concerned, the dramatic increase in the level of imports from Colombia is of particular note. Colombia attracted much attention in the Parliamentary debate on Wednesday 21 October 1992 when it was divulged that Leeds city council had signed a letter of intent with the Colombians to import coal to meet the city's needs. This was forced on the council by the government requirement to open all contracts to competitive tendering and led to calls that the contract should be cancelled because the Colombians use 'slave labour' in their mines. Nevertheless, Britain is likely to continue to import cheap Colombian coal.

Table 9.7 Imports into the UK of coking coal, 1983−91

	1983	1984	1985	1986	1987	1988	1989	1990	1991
Australia	1121	1648	2988	2274	2668	2326	2204	2966	4044
Canada	0	109	337	423	362	518	720	961	743
Germany	0	1	0	1	0	1	16	2	3
USA	908	2944	2647	2433	2502	3058	3864	3791	4107
Other	0	29	14	2	8	38	27	7	3
Total OECD	2029	4731	5986	5133	5540	5941	6831	7727	8900
CSFR*	0	0	3	0	0	0	0	0	0
Poland	414	740	1160	1097	952	1137	975	794	238
CIS	0	0	0	27	0	4	19	93	0
South Africa	0	0	1	0	0	0	0	0	1
Colombia	0	0	0	0	0	0	68	0	0
Other	1	1	0	0	0	0	105	0	1
Total non-OECD	415	741	1164	1124	952	1141	1167	887	240
TOTAL	2444	5472	7150	6257	6492	7082	7998	8614	9140

* Czechoslovak Federal Republic.
Source: OECD.

Table 9.8 Imports into the UK of steaming coal, 1983−91

	1983	1984	1985	1986	1987	1988	1989	1990	1991
Australia	866	211	2469	1548	791	1090	472	67	610
Canada	1	8	61	26	0	36	9	24	17
Germany	523	1033	783	461	261	190	271	248	217
USA	234	400	498	1126	388	1014	1011	1641	4025
Other	180	801	525	429	937	472	387	1868	312
Total OECD	1804	2453	4336	3590	2377	2802	2150	3848	5181
CSFR	0	33	10	0	0	6	4	2	5
Poland	95	619	258	120	132	133	80	250	319
CIS	23	11	23	21	89	371	163	499	900
China	8	22	34	26	146	319	450	69	294
Indonesia	0	0	0	0	0	0	16	0	97
South Africa	58	257	724	312	188	276	346	356	577
Colombia	0	0	89	56	319	574	806	956	2447
Venezuela	0	0	0	0	0	92	86	169	230
Other	24	27	107	171	36	25	34	19	294
Total non-OECD	208	969	1245	706	910	1796	1985	2320	5163
TOTAL	2012	3422	5581	4296	3287	4598	4135	6168	10344

Source: OECD.

Chapter

10

Supply/demand and prices

THE THREE COAL MARKETS

There are essentially three main types of demand for coal. The first of these is the nascent home demand, whether it is for the basic steaming or for the more specialized coking coal. The reason why both types of coal have to be combined in a local market analysis is because of the localized incentives and restrictions on competition in many areas of the world. Second and third are the export markets for coal and, as the demand for the two types of coal varies around the world, they are effectively two specific areas of demand. This difference is reflected in the higher price which coking coal generally fetches both internationally and in home markets.

Import restrictions are related to a government's desire to support its local mining industry and to the difficulty of import substitution, often resulting from the geographical position of the country. A third, and increasingly important, factor is the effect which import substitution will have on a country's balance of payments. If a country needs to import oil and gas it is more likely to attempt to protect an indigenous coal industry in order to provide a source of energy with no exchange rate risk. Indeed, not only did Sir Derek Ezra, a past Chairman of the NCB (and since Lord Ezra), argue in defence of the coal industry that 'coal imports on any scale would represent a significant burden on the balance of payments, at a time when the balance of payments benefits from the North Sea could be declining' (*The Energy Debate*), he also stated that 'the price of imports into the UK is heavily influenced by the sterling exchange rate, the long term level of which is problematical'.

The geographical situation of the country is one area which originally provided the UK with a great advantage as British coal could be sent all around the world in British ships and was stored in various depots to fuel those ships. It is now a liability as the same ports which exported cheap British coal are able to import still cheaper foreign material. Conversely, a country like Botswana, which has extensive coal reserves, may find it impossible to exploit them economically because of the country's landlocked position which is likely to restrict the profitable export of the potential coal production.

The local market for coal is, therefore, often a false market and bears no relation to the international or free market price of the commodity. Nevertheless, it must be true to state that the price of coal should rise in relation to the distance from the mine as the extra transport costs should lead to an increase in the delivered cost of the product. On top of this must be placed an 'in use' charge which takes account of the additional cost of transporting coal over a distance of, say, 1000 miles when two 500 mile deliveries could be made in the same period of time. The method of transport from the source is also important, as the need to import coal during the 1984/5 miners' strike led MacGregor to discover that 'it cost more to transport coal by rail from the docks at Hunterston to Ravenscraig a few miles away than it would to ship that coal half the way from America to Hunterston'. Therefore the price of coal will rise depending on the distance and method of transport from the source.

If, however, a coal producer can extract coal at a price which is significantly lower than the production cost in a potential export market then it may be economic to attempt to break into that market. The local market may, also, attempt to import coal if it calculates that imported coal would be cheaper than that produced from local mines. As such these factors tend to distort what may be called the international price of coal and any CIF price cannot realistically be compared with another such price when considering the international market unless it is in relation to a particular and well defined area. In this instance a CIF price would show the delivered price of the material in one particular area of the world and give an indication of the relative cost of coal from all sources in that one particular market. It could not be used to indicate that the same coals would be relatively as expensive or cheap in any other market.

For a true picture of the international market price of coal an FOB price

Table 10.1 FOB and CIF prices of steam coal delivered into Europe in 1986

Country	Production type	FOB cost (US$/t)	Transport cost (US$/t)*	Delivered cost (US$/t)
South Africa	Open cast	13.50	7.49	20.99
	Underground	19.00	7.49	26.49
	Richards Bay Phase IV open cast	21.00	7.49	28.49
Australia	Queensland open cast	22.00	12.32	34.32
	New South Wales open cast	30.50	12.32	42.82
	New South Wales underground	31.00	12.32	43.32
Colombia	Open cast	50.00	5.95	55.95
USA	Large open cast	40.50	5.94	46.44
	Large underground	42.00	5.94	47.94

*Assumes US$25/barrel oil price in working out freight rates.
Source: House of Commons Energy Committee Report on the Coal Industry, 1986/7 session, HMSO.

has to be used with any additional transport cost added at the ruling cost per tonne per mile travelled. This additional cost can then be compared internationally and the most expensive or cheap coal in any one area can be ascertained. Obviously companies tend to keep the FOB prices as discreet as possible because of their preference to sell material CIF, which allows an export margin to be added and allows for the use of the ship on the return journey where additional profits could potentially be made. The FOB and CIF prices of steam coal delivered into Europe in 1986 are shown in Table 10.1.

Basic freight rates for coal to Europe in 1986 are shown Table 10.2. It will be seen that these rates differ from the total transport costs shown in Table 10.1 as they are the basic freight rates and do not take account of the additional insurance costs, nor do they leave room for a profit margin for the supplier. As can be seen it would normally be cheaper to purchase coal FOB and to carry the additional transport costs rather than to buy coal on a CIF basis.

The other factor, of course, which affects the international price of coal is the relative supply demand position of the product. The international market in coal amounted to some 401 mt in 1991. This represents only some 9.0% of

Table 10.2 Basic freight rates for coal to Europe in 1986

Country of origin	Freight rate (US$/t)	Implied insurance and profit (US$/t)
South Africa	4.25	3.24
Australia	6.00	6.32
Colombia	2.55	3.40
USA	3.87	2.07

Source: House of Commons Energy Committee Report on the Coal Industry, 1986/7 session, HMSO.

total world production of coal in 1991 and shows clearly how small the international market is in relation to the use of coal in its home markets. This is most importantly related to additional transport costs but, as mentioned above, is also related to import restrictions and local subsidies. In view of this relatively restricted market, and a local producer's natural preference to sell into the local market because of the much higher margins which can normally be achieved in comparison with exports, the international coal market can experience feasts and famines of material and the price of coal can fluctuate quite widely. This is particularly the case when there is a world shortage of energy supply as occurred at the beginning of the 1980s. At these times coal prices rocketed (Figure 10.1).

Finally, whilst the specific difference between coking and steaming coal

10.1 Coal prices CIF Europe 1977–91 (US$/tonne)
Source: OECD.

prices was mentioned above, each coal from each different source commands a different price, particularly on the international market. This is simply because of the different composition of each type of coal. Different coals produce different amounts of energy in relation to their specific chemical composition and the relative amount of carbon they contain in proportion to the other material. This other material can include the sands and other non-combustible matter which were deposited at the time of formation of the original swamps from which coal was formed, or can include sulphur which was a later arrival as fluids percolated through the rotting and compressing matter (see Chapter 8). As a result some coal contracts (such as those entered into between British Coal and the electricity generators) are calculated on the basis of the energy content of the coal (gigajoules/tonne) rather than its tonnage. It is then up to the customer and supplier to determine the energy output of the coal so that the number of tonnes to be delivered can be determined. British Coal uses a base of 23.8 GJ/t for steaming coal, 26.6 GJ/t for industrial coal and 29.7 GJ/t for domestic coal.

The difference in composition is most clearly shown by the higher price which is awarded to coking coals, although these coals also have to be analysed to determine their own chemical composition. This is because the use to which they are put is often more chemically dependent than that simply for heat production for the generation of electricity. Specifically, the use of low sulphur coals is necessary in the manufacture of iron as shown by Abraham Darby at Coalbrookdale in 1709 (see page 10). His great advantage was that locally produced coals had a low sulphur content and therefore were particularly suited to his process. The coal at this location has now been worked out and the technology for the manufacture of iron has advanced considerably with in-furnace chemical analysis. Nevertheless, the chemical composition of coal remains an important consideration in demand.

It is slowly becoming evident in the steam coal market that low sulphur coals are starting to command premium prices over high sulphur material. This is not easily determinable as far as the pricing of coal on the international market is concerned as most internationally traded coals are of low sulphur and low ash content because they would not otherwise be in demand. However, the increasingly stringent emission control legislation which has been introduced in the USA has led to higher demand and prices for low sulphur coals. One of the innovations made in the USA has been to introduce a pollution allowance for all power generators which allows them strict limits on the amount of sulphur they can produce in any one year. Those generating utilities which purchase low sulphur coal and thereby stay within their own emission restrictions are able to sell their additional allowance to other generators which have contracts to purchase high sulphur coal and can not afford to retrofit FGD

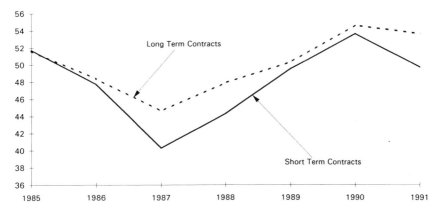

10.2 Short term and long term coal contract prices CIF Europe 1985–91 (US$/tonne)
Source: OECD.

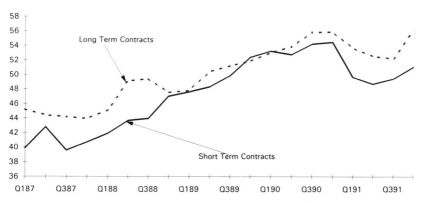

10.3 Quarterly short term and long term coal contract prices CIF Europe (US$/tonne)
Source: OECD.

equipment. This legislation ensures that the USA as a whole will not exceed a set sulphur emission level, although it will mean that some areas of the country will be significantly cleaner than others.

Finally, it should be remembered that the guarantee of long term supply commands a higher price for the product. This is one of the problems the British coal mining industry suffered as a result of the 1972, 1974 and 1984/5 strikes. As long term supply could not be guaranteed, because of the perceived militancy of the mining unions, consumers looked elsewhere for alternative supplies and were not won back by British Coal following the resolution of these strikes. The higher price that this guarantee of the long term supply commands can be seen from Figure 10.2, which compares the prices paid for coal in Europe on the basis of both short term and longer term contracts. From this it can clearly be seen that longer term contracts command higher prices. Figure 10.3 shows recent prices on a quarterly basis, from which it can be seen that there is some, but overall not much, seasonality in coal prices as much of the internationally traded coal is coking coal which will not be affected by higher demand from electricity consumers during cold weather. This is vastly different from the seasonality that coal prices exhibited before the twentieth century and shows the development of local stock-handling and transport systems.

FUTURE SUPPLY/DEMAND

As far as the future supply of coal is concerned, the developing countries are likely to become more important in the export market. This is simply because these countries have low labour costs and, unfortunately, the need for work forces the population into employment which may not be as safe as that available in more advanced economies. For instance, coal producers in both the UK and Australia give clear indications about the mortality rates of their employees in the mining industry but these details are not readily apparent in the developing countries.

Whilst not strictly a developing country or suffering from a stigma associated with high mortality rates South Africa has shown that it is always possible to circumvent sanctions, albeit at a price. In the case of South Africa trade sanctions were imposed in order to pressure the government into the abolition of apartheid but South African coal producers were still able to sell a large proportion of their coal into the Amsterdam/Rotterdam/Antwerp spot market. This meant that the final consumer did not necessarily know the origin of the coal which had been purchased. As a result South African coal exports were not severely affected by the imposition of sanctions although a discount had to be applied in order to encourage buyers. The lifting of sanctions has led to a rise in the price of South African coal and the consequent desire of the producers to expand their export markets has forced them to consider the expansion of the Richards Bay coal terminal. The rise in the price of South African output has not been as great as may have, initially, been expected following the removal of the apartheid discount. This is because the market has moved to an equilibrium some way between the higher prices asked for other internationally traded production and the discounted South African output.

Whilst the UK production of coal was forecast in 1992 to decline by 35 mt over the next two years, it is unlikely that there will be a corresponding 35 mt increase in the level of British coal imports. This is largely because most of the coal is used to generate electricity and will be replaced by newly commissioned gas-fired power stations rather than the country's future electricity needs still being met by coal- or nuclear-generated electricity. In any case the fall in production represents only 0.8% of the world's total output in 1991 but a somewhat more significant 8.7% of internationally traded coal in 1991. If British demand for coal from the international market is increased by the full amount of the proposed cut then the UK will account for some 13.6% of total world coal demand. This would be likely to have some effect on international coal prices were it not for the fact that other coal producers are attempting to increase their production specifically for the export market (see Chapter 8).

In the future, therefore, there is likely to be sufficient coal to meet international demand unless there is a great energy squeeze, as has occurred twice in the last 20 years. As the expansion of the world economy continues and the less developed countries become industrialized, there is likely to be an increase in the demand for energy and the value of energy is likely to rise. Even if the world is able to banish inflationary pressures the price of energy is likely to rise and, unless a non-polluting and cheap alternative is found, the price of coal will also rise as it will be carried in the wake of the potential rise in the prices of oil and gas.

Based on the information in Chapters 8 and 9 the following international coal supply/demand balance can be calculated (see Table 10.3). It should be noted that the shortfall in 1992 and 1993 will be met from stocks (if demand turns out to be as strong as projected). As can be seen the potential future supply of coal on to the international market will more than meet demand and it may be that some of the start-ups in Latin America are delayed due to excess capacity elsewhere in the world. Under this scenario the world coal price is expected to remain weak for some considerable time although a

Table 10.3 International coal supply/demand balance (mt)

	1991	1992	1993	1994	1995	2000	2010
Supply	400.9	398	408	423	444	520	612
Demand	396.0	404	412	420	428	473	575
Balance	4.9	−6	−4	3	16	47	37

potential rise in the oil price, due to increasing scarcity over the longer term or as a result of heightened Middle Eastern tensions over the short term, would undermine this assumption.

BRITISH COAL PRICES

The price paid for coal in the UK was, until recently, a price which was set between British Coal and the electricity generating company, the CEGB. Following the privatization of the CEGB in March 1991, when two separate generators, PowerGen and National Power, were created, British Coal has continued to supply coal to the generators on a contract which expired at the end of March 1993. Following the expiry of the contract it is anticipated that the price paid by the generators will fall from around £1.88/GJ to £1.51/GJ in 1993/4 and then to only £1.33/GJ in 1994/5. This represents a fall of some 28% and is the main reason behind the government's belief that the amount of British coal for which there is a market will decline because import substitution at a price of around £1.30/GJ is forcing this decline.

Intriguingly, despite the production squeeze resulting from the debilitating effects of the coal strike, 1984 was the only year in the past nine when home-produced steaming coal was cheaper on an average CIF basis than the imported version. This was caused by the relative strength of the dollar making the dollar converted cost of British coal cheaper rather than there being any fall in the price of international coal. Indeed, the international price of coal imported into the UK actually rose in 1984 although the overall price of the commodity on a CIF Europe basis fell. Admittedly there were some imports which were priced below the average price but the fluctuation in some import prices seems extraordinary as shown in Tables 10.4 and 10.5. Some of the rise in 1984/5 was obviously related to the miners' strike and the need to import coal to ensure that industry had sufficient supplies. If the country becomes more dependent upon imports for electricity generation then a rise in import prices would have a much more significant effect on the cost of electricity. The fluctuations in the exchange rate also make a large difference to the cost of imported coal, as it is priced in US dollars, and the cost of imports will rise as sterling falls, thereby exacerbating an already weak trading account.

Tables 10.3 and 10.4 should be compared with the tables of UK coal imports in Chapter 9 (Tables 9.7 and 9.8) in order to obtain a full understanding of some of the numbers. In particular some of the fluctuations in prices of the coal imports (for instance the Australian increase in steam coal

Table 10.4 Fluctuations in the import price of coking coal, 1983–91 (US$/t CIF)

	1983	1984	1985	1986	1987	1988	1989	1990	1991
Australia	59.37	58.51	60.50	55.32	53.10	50.32	50.37	76.38	59.74
Canada	—	62.29	64.53	53.69	52.84	53.79	56.87	67.80	60.32
USA	64.97	60.76	64.17	59.92	60.99	60.72	59.90	72.47	64.43
Poland	54.46	54.08	64.68	57.81	58.15	57.83	57.16	65.78	59.70
CIS	—	—	—	58.82	—	44.17	50.42	46.99	—
AVERAGE	60.46	59.20	62.74	57.45	56.87	56.49	56.66	72.40	61.89

Source: OECD.

Table 10.5 Fluctuations in the import price of steaming coal, 1983–91 (US$/t CIF)

	1983	1984	1985	1986	1987	1988	1989	1990	1991
Australia	70.64	67.63	60.26	59.34	58.75	68.14	68.75	96.29	50.51
Canada	—	68.12	54.60	51.54	—	55.65	57.72	58.76	—
USA	58.27	76.66	70.73	50.00	48.26	70.73	66.32	68.83	60.26
Czechoslovakia	—	75.83	60.72	—	—	89.79	—	102.72	89.50
Poland	71.86	83.67	88.27	77.73	88.01	91.02	70.81	84.49	100.64
CIS	72.86	90.71	62.79	36.38	38.55	47.52	56.16	53.29	49.16
China	—	—	—	48.89	66.42	41.18	60.22	71.72	60.98
South Africa	45.85	61.04	53.45	45.22	44.68	57.83	48.40	—	59.13
Colombia	—	—	73.89	54.21	62.48	72.21	65.04	72.96	57.62
AVERAGE	66.59	74.02	62.04	55.32	57.24	66.51	64.03	68.92	59.24

Source: OECD.

costs in 1990) are due to a relatively small amount of coal imported. In this case the fixed overheads and other costs of importing the coal are excessive and swamp the underlying cost of the coal. Even without the massive rise in Australian coal prices in 1990 the price fell back below the 1989 level in 1991 and was significantly cheaper than all other imported coals in 1991 except for CIS steaming coal. Given the CIS' need to maximize its export earnings the low level of export prices is understandable. However, it may not continue into the future as the coal mines are modernized and receive foreign investment which requires a positive return on capital.

In the future the UK's increasing dependence on imported coal is likely to lead to a reduction in central and other fixed overhead costs but this may be more than offset by the increase in demand for coal which could so squeeze the market as to put upward pressure on the coal price. In addition the world is currently in the midst of the most serious economic depression since the 1930s. Whilst the international coal trade expanded between 1990 and 1991 the actual price of the commodity dropped and the price of coal may well remain weak for some time to come, especially given the large stockpiles around the world. As the world economies start to recover, assuming that they do, then the stockpiles of coal are likely to deteriorate and the price should begin to recover as coal becomes more scarce. This rise will be particularly acute if the rises in oil and gas prices, which are likely to precede a rise in the coal price, encourage fossil fuel consumers with the ability to switch between energy sources to move to consume coal in preference to oil or gas.

This competitive market in energy supplies is demonstrated in Figures 10.4 and 10.5 which show the relative costs of electricity generation in sterling and in US dollars per tonne of coal equivalent. These show that the price of coal used in electricity generation in the UK remained relatively stable between 1983 and 1991, and was, at times, helped by fluctuations in the exchange rate. Nevertheless, the declining cost of heavy fuel oil has made coal much less competitive in the cogeneration market although natural gas has remained more expensive on a tonnes of coal equivalent basis. In fact the cheapest form of electricity generation for this country comes from imported coal, with CIS coal at US$49.16/t (equivalent to around £28/t) the cheapest. Obviously the security of CIS supplies may be brought into question but even so the low cost of Australian imports provides a more secure alternative.

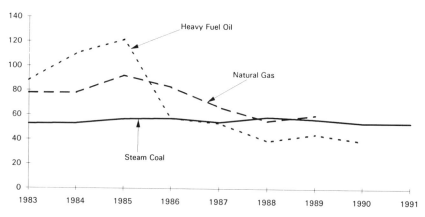

10.4 Electricity generation costs 1983–91 (£/tonne coal equivalent)
Source: OECD.

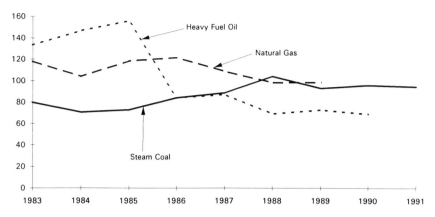

10.5 Electricity generation costs 1983–91 (US$/tonne coal equivalent)
Source: OECD.

These currently low prices are, however, related to the world energy glut which has occurred as a result of the recession. Coal, gas and oil producers are selling their output in order to generate cash flow rather than just to generate profits, and the situation could change drastically when the world economy starts to recover. When this occurs the Australians are likely to find that the Japanese market becomes much more profitable for them as the prices offered by Japanese companies return to pre-recession levels. As such the price of exports to the UK is likely to rise and reduce the presently competitive nature of the Australian output. In such a case the competitiveness of internally produced coal is likely to increase as overseas energy producers find that their output is in demand and they raise their prices accordingly.

It is in such situations that a premium price for the stability of long term supply becomes important. Almost all consumers forget about the bad times when they are in the middle of a boom – witness the dire situation which exists in the UK at the moment in comparison with the situation which existed in the country during the heady days of 1987 and the even headier days of the post-crash low interest rate inspired boom of 1990. At that time there was no worry that interest rates would ever rise or that property prices would ever fall. The same is just as true of the coal market today.

At some stage in the future the price of coal will rise and only those electricity generators which have ensured their supply on long term contracts will be able to continue to purchase coal effectively and economically. As such, the simple argument that gas is cheaper than coal today will not stand up to analysis and this is one reason why all the new gas-fired generators are purchasing their gas supplies on long term contracts (see Chapter 14).

However, the future supply of coal needs to be guaranteed by the electricity generators and one method by which they can achieve this is to enter into long term contracts with their local, indigenous coal producer. That the price of the contract is seen to be too high over the short term belies one of the stipulations in the electricity privatization Act which set up the Office of Electricity Regulation (Offer). This stipulation is that the generators have to provide electricity at the 'best effective price reasonably obtainable, having regard to the sources available' and an effective price can be taken to mean a price which guarantees long term supply.

The dependence of the UK for its energy supplies on imported coal and on gas which may only last a relatively short period of time and could be exported if international prices are higher, means that the cost of energy under the proposed regime is likely to fluctuate more widely in the future than it does at present. This can be expected to reduce British competitiveness on the international market as energy consumers will find it ever more difficult to price their products and will have to include higher margins to take account of this uncertainty. It is no wonder that in the industrial heartland of the country in the eighteenth and nineteenth centuries industrialists set up their operations close to the source of power, be it fast running streams or coal for steam generation. And in order to guarantee their supplies of power further they dug their own coal mines or bought stakes in others and, whilst the unions fought this because it could lead to transfer pricing whereby the coal mines' profits were reduced and the main part of the profit taken in the manufacturing company, at least it guaranteed the consumer a source of energy, and the worker continued employment. Of course, this is effectively what the RECs are doing at the moment but it begs the question of what will happen when the gas runs out and there is insufficient finance to cover the cost of a new coal mine.

The simple answer is to let the free market decide as, by the time the supply of gas does start to decline, energy consumers will be aware of the problem and will have to investigate alternative sources of power. If they still require electricity then the cost of that electricity will rise as the gas price rises to take account of the depletion in reserves. And, assuming that no other, cheaper sources of power are discovered in the meantime it is sensible to anticipate that the rise in energy prices generally will enable the financing of a new mine if this is required by a rise in imported coal costs. Indeed, when the gas runs out Britain could be in an even worse state than today if it was forced to purchase ridiculously high cost coal from state owned mines rather than being allowed to import coal from much cheaper producers overseas.

However, it must be the consumer who decides whether or not to accept a rise in energy costs, as was the case in the mid-1930s when ICI, among other companies, offered to pay higher prices for coal if the whole of the increase in price was passed on to the miners who were at that stage threatening industrial action in pursuit of higher pay. ICI, and not the colliery owners, was effectively dictating its own power costs and ensuring its future supply. The pressures of international economics are now such that this situation is not likely to recur. ICI has recently closed two of its chemical plants stating that energy prices are too high, whilst the Major Energy Users Council has instituted a campaign to attempt to reduce the cost of electrical power to its members.

Part
III

Privatization

The Thatcher legacy

THE ULTIMATE PRIVATIZATION

Cecil Parkinson, a close ally of Margaret Thatcher, dubbed the privatization of the coal mining industry 'the ultimate privatization' when Secretary of State at the Department of Trade and Industry (DTI). However, the DTI curse affected him as much as it has affected all of the Ministers at the Department since the 1979 general election, and he was forced into resignation after only four months in office. Michael Heseltine, despite changing the title of the senior appointment at the Ministry to the President of the Board of Trade, in an attempt to dispel the curse, also found his first year in the job far from easy. However, the White Paper on the future of the coal industry seems to have extricated him from the immediate problems which surrounded the October 1992 pit closure announcement.

Mrs Thatcher, a free marketeer through and through, placed privatization high on her political agenda. Early in her tenure at No 10 Downing Street it was made clear that the coal industry was included and, strange as it may seem today, the initial plans suggested that privatization would take place in 1983. However, the plans were thwarted by the inability of the NCB management to return the industry to profit, partly as a consequence of the failure to proceed with the 50 pits over five years closure programme put forward in 1981. Then, in April 1982 Arthur Scargill had succeeded the more moderate Joe Gormley as President of the NUM, and it became clear that the return of the industry to profit would probably lead to a long and expensive strike. For, after he had met Arthur Scargill officially for the first time Nigel Lawson said that 'there was no way we could do business with him'. As a result government strategy was geared to preparing for a strike in order to ensure that the perceived humiliation of the Conservative Party in 1974 did not recur.

Two of the legacies the industry now faces are the intransigence of both Margaret Thatcher and Arthur Scargill during the strike and the large amount of misdirected investment in the industry at the time. A lack of capital expenditure during the 1984/5 strike and a general failure to spend on productivity enhancing programmes was one of the reasons cited for the need to close so many pits in the 1992/3 closures. This meant that British Coal was not internationally competitive and that the electricity generators would be forced by their regulator to buy cheap foreign imports at the expense of locally produced coal. In one respect this is false logic because the investment which has taken place since the end of the strike will now improve the profitability of the industry and thereby make it more competitive on an international basis. And it would still have been necessary to have carried out the investment in any case because of the decline in output which would otherwise have resulted (see Table 11.1).

However, the investment in new faces to produce more coal is also a false guarantee of the future of the industry if it occurs at the expense of truly

Table 11.1 Predicted coal output with and without investment up to 2000/1

Year	Output without investment (mt)	Output with investment (mt)
1992/3	65.0	65.0
1993/4	65.0	65.0
1994/5	59.6	62.2
1995/6	54.2	60.0
1996/7	42.8	57.8
1997/8	41.6	56.9
1998/9	41.6	54.7
1999/2000	36.1	53.4
2000/1	19.8	47.7

Source: British Association of Colliery Management.

productivity enhancing programmes. Indeed, it seems that the NCB continued to spend vast sums of money on new faces even though it would be unlikely to attain the levels of productivity necessary to repay its capital and interest costs; let alone provide a level of profit acceptable to the private sector. Unless, of course the Corporation continued to receive high prices for its output, in which case it could never be privatized. In this way the management of the NCB/British Coal seems to have thwarted the government's attempt to privatize the industry and the government must have come to believe that the industry would have to be forced to comply with its ideologies. Indeed, having spent vast sums of the taxpayer's money in order to win the coal strike British Coal failed to follow through with the changes in working practices that winning the dispute should have allowed. As a result it is probable that more pits now have to be closed than might otherwise have been necessary.

That the labour problems in the industry needed to be sorted out is without question, and that they needed to be sorted out before new investment could be justified on economic grounds is also without question. However, it must be seen to be still worse management if those same miners who were defeated in the strike should be penalized before they have been given a chance to show the potential benefits of the investment which has taken place since the end of the strike. And for the government to show the same lack of respect for the UDM miners cannot be justified if the government is not willing to allow those miners who have already proved that they want to work (often in the face of violent opposition) the chance to show that they can do it profitably. In this respect the decision to sell off those mines that British Coal does not want to retain must be applauded and British Coal must not be allowed to put up any barriers to entry as this will distort the free market in the coal sector still further.

Additionally, the continuing wish to penalize British Coal for as long as possible by charging market interest rates on its debt is completely unrealistic as it has hindered the Corporation's investment programme to the advantage of the Treasury. This is the same Treasury which could be penalized for not allowing sufficient investment in the coal industry as, despite offering the pits for sale, it still remains up to potential investors to decide whether or not to buy any of the twelve pits on the closure list.

DREAMS AND DESIRES

Politics plays a major part in any strategy adopted by the party in office at any one time. There is a need both to keep the electorate happy today, and, more importantly, to attempt to make them happy at the time of the next general election. Owing to people's inherent fear of change any changes have to be pursued slowly, unless the government is convinced that the change will be for the better. It must have been clear that the desire to privatize the coal industry in 1983 was a dream inspired by the wish to remove coal mining from the political agenda. This, clearly, proved impossible and so the government embarked on the long path to create a truly profitable coal industry rather than one which made profits by virtue of charging unsustainably high prices.

And, despite the threatened strike in 1981 and the year-long conflict in 1984/5, which caused some to question the need to create a profitable industry, the population has largely agreed with the government's strategy. This is because it was made clear that the money saved by a reduction in the subsidies which were given to the nationalized industries would be available for tax cuts. Obviously, by the start of the 1990s the public's changing ideals meant that the logic of sacking miners to add to the unemployment statistics in the depths of the worst recession since the 1930s was more difficult to find. However, the bulk of the industry had been reduced to a seemingly profitable core and the electorate was faced with a straightforward question: either everyone paid an extra subsidy on top of their electricity bills in order to keep the pits open (on top of the one already paid for the decommissioning of the old Magnox nuclear power stations) or the pits would have to close. The mood swing in the country against a subsidy became more apparent when there was an uproar over the imposition of VAT on domestic fuel supplies in the March 1993 Budget. This showed that, whilst the public sympathized with the miners, personal costs outweigh political ideals.

In furthering its desires for the coal industry the government has been aided, albeit with some reservations, by many public and private bodies. Probably the most august of these was the House of Commons Energy Select Committee which was disbanded following the closure of the Department of Energy and its amalgamation into the DTI. This Committee produced a report on the Coal Industry which was printed on 28 January 1987 when the strike was still at the forefront of people's minds. Against the obvious need for a profitable industry the government's brief was that its 'first energy policy objective . . . [was] that there should be adequate and secure supplies of energy available to the people of this country', and that, as a result, 'coal will be vital to the country's energy needs'. The report continued that 'indifference to the future of the British coal industry would be an abrogation of political responsibility, and we are pleased that the Secretary of State recognises the strategic need for a British coal industry'. However, the report was written on the basis of an inaccurate assumption – namely that there would not be an economic alternative to the production of electricity from coal. And, although the Committee admitted that overseas competition would lead to 'a persistent pressure on the real price of British coal . . . for the foreseeable future', it failed to realize just how devastating this would be.

The unachieved desire to create a smaller and profitable, but still relatively large, coal industry was one of the failures of the Thatcher years. The government spent much time thinking about it but, in the end, the dream faded as the years, and other over-riding priorities took its place. One of the reasons behind this must be related to the militancy of the mining unions, for the need to reduce manpower to improve productivity would almost certainly have led to more industrial action. However, the government must also be partly to blame for these failures as it failed to repeal some of the legislation which meant that some of the inefficient working practices had to be maintained despite technological advances reducing the need for them. The government was able to withstand the 1984/5 strike by virtue of strategic planning but, despite the fragmentation that would have occurred had the industry been returned to private ownership earlier, it is unlikely that independent producers would have been able to survive a year-long stoppage.

Following the strike the government's privatization bandwagon kept the Department of Energy busy − with the sales of further shares in BP and the new issues of all of the electricity stocks. Coal, which should have been put at the top of the list, was pushed to the bottom and, in order to make the electricity stocks more attractive it was only offered one three-year supply contract rather than a raft of different contracts with different prices and expiry dates. This would have ensured both a steady improvement for the electricity companies as new lower priced contracts were negotiated on a rolling basis, and a longer term future for British Coal which would have given it set targets for which to plan. Indeed, a range of long term contracts would have made British Coal much more saleable to the private sector as potential buyers would have seen the possibility of improving productivity quickly and thereby increasing the return on the short-term higher priced contracts whilst still having the security of longer term contracts. As it is Mrs Thatcher's privatization dream turned into John Major's nightmare − and it could still come back to haunt him in the future.

HOW DID WE GET HERE?

The last 13 years have seen considerable changes in the way that government is seen to operate throughout the country and the responsibilities which it is supposed to have. The most significant of these is its desire for privatization as a method of reducing the State's involvement in many areas of industry and commerce which are not seen to be part of its role in a modern, and more importantly, free market, society. Much of this is plain to see, however, the speed of change has, in some instances brought up a whirlwind and caused significant problems. This whirlwind was largely caused by the need for speed brought about by the British political system with elections held at least every five years. This means that as much in the way of privatization as possible had to be completed in every parliamentary term so that it would be more difficult for a succeeding opposition party to reverse many of the privatizations which had taken place. Note the yo-yo damage done to British Steel through its periods of national and private ownership since the Second World War.

The reversal of many policies was for a long time part of the Labour Party's manifesto and it still intends to re-nationalize some of the industries that the Conservatives have placed in private ownership. Even the Labour Party, however, has fallen partly under the spell of free market economists, although the lack of finances to proceed with much in the way of nationalization must be partly behind this changing stance – especially because of the difficulty experienced by the Conservatives in running a balanced budget. As mentioned before, the need to push ahead with the privatization programme in order to finance the tax cuts which brought the Conservative Party to power in 1979, and returned them in a surprising fourth election victory in 1992, caused many financial upsets.

This meant that if one industry was not ready for privatization then another would be put in its place to take advantage of the boom, and may have meant that the Treasury received less from the sales than could have been achieved had they been made at a more propitious time. Nevertheless, the government did have some successes, although these must, at least in part, have been inspired by the stock market boom in the run-up to the October 1987 crash and by its recovery ensured by the dramatic interest rate reductions in late 1987 and early 1988. This is as much a part of the free market as anything, and is a policy followed by Lord Hanson in his eponymous company, as well as by the government (note the flotation of SCM in the USA and the scathing reports now afforded Hanson on what was an astute deal, and compare it with the lack of damnation in much of the Conservative leaning press on the poor performance of an investment in British Steel shares since the time of the flotation – both are the result of free market economics).

Following the 1984/5 strike it was anticipated that British Coal would return to profit and be in a fit state for privatization towards the end of the decade. The clear inability of the industry to meet these targets forced the government into a rethink on the privatization issue and pushed the potential date for it into the early 1990s. This meant that the power companies, in the shape of the electricity generators and distributors, were brought in to be privatized ahead of the industry which depended upon them for support. Even the electricity generators were allowed to be altered for privatization because the government was unable to persuade the institutions that future profits would more than cover the costs of decommissioning the old Magnox nuclear power stations.

As a result, the nuclear element of the industry was left in government hands and the much larger National Power was left at its old size rather than being split into smaller operating units which, together with a possible split of the smaller PowerGen, would have increased competition. Unfortunately, this did not occur because of the pressures of the privatization timetable. This left British Coal with one major consumer and one smaller consumer, PowerGen, and meant that the electricity distribution companies did not have a broad enough selection of companies from which to buy power to enable the creation of what could be more fairly termed a free market. Even in its originally planned form, with the nuclear power stations retained within National Power, there would not have been a broad enough selection of generators to guarantee a free market in electricity.

The slippage and delays which have been caused have meant that the coal industry is now to be privatized to absolve the government of previous mistakes. Putting the 12 pits on the closure list up for sale is no recompense for the failure to secure a range of coal supply contracts in the run-up to electricity privatization (see above). And even the new five year contracts provide little in the way of long term guarantees for the 32 pits that will remain in production or the 6 to be mothballed. However, whilst they will be difficult to live with, the new contracts (£1.51/GJ declining to £1.33/GJ against 1992/3's £1.88/GJ) will increase pressure on British Coal to become a more productive, and hence lower cost producer. Unfortunately, this pressure meant that British Coal had to announce the closure of so many pits in October 1992 as it was unable to improve productivity rapidly enough in the time allowed. In part this must be related to the intransigence of Arthur Scargill and the NUM's reluctance to accept new working practices but a large part of the difficulties is also related to the blinkered desire of the Conservative government, under Margaret Thatcher, to balance its books through the privatization of the electricity industry with little thought about the long term competitive problems which would reduce the demand for coal-generated electricity.

BRITISH COAL FIGHTS BACK

The Thatcher years were not always a bed of roses in the Conservative Party's relationship with the management of the British mining industry. There were a number of occasions when British Coal personnel, whether individually or collectively, were accused of leaking documents to the press because they did not agree with the strategy which had been forced on the Corporation in the run-up to privatization. Indeed, in his speech to the House of Commons in the debate on the coal closures on 21 October 1992, David Howell, the Energy Minister at the time of the 1981 confrontation, accused British Coal of leaking details it did not like and leaving him 'with a fizzing bomb in his hand'. Michael Heseltine was also concerned about the leaks about planned closures in 1992 and he stated that this was one reason why the announcement of the pit closures had been brought forward and made before Parliament had reconvened from the summer recess.

Despite these problems, British Coal can be clearly shown to have moved along the government's chosen path in pursuit of privatization. This can be seen from the improvement in the overall productivity of the mining operations in the first four years of the Conservative administration. Indeed, under the previous Labour administration, which introduced the 1974 Plan for Coal, the productivity of the coal mines had fallen both in terms of output per manyear and output per manshift from 474 t and 2.29 t to 448 t and 2.24 t respectively. Between 1979 and 1983, i.e. before the start of the miners' overtime ban on 31 October 1983, productivity had improved to 504 t/manyear and 2.44 t/manshift (Figure 11.1).

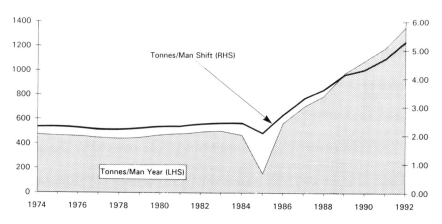

11.1 Productivity at British coal mines 1974–92

A large part of the increase in productivity must, however, be related to the long term nature of the improvements wrought by the 1974 Plan for Coal in view of the long term nature of the investment needed to increase output in the mining industry. Indeed, as Derek Ezra wrote in 1982, 'there is a gap of up to a decade before full benefits of major investment are realized so that the industry is now seen to be significantly influenced by the under-investment of the 1960s' (*The Energy Debate*). Within the framework of the Plan for Coal the number of deep mines run by British Coal fell from 246 to 191 between 1974 and 1983, and this must have had an impact on the productivity improvement of the industry. Additionally, the Plan for Coal led to the expenditure of some £6.5 billion on new mines which had a total new capacity of some 42 mt, particularly at the massive Selby complex which is now likely to dominate the rump of the industry.

Some of the 1974/9 Labour government's commitment to the coal industry was related to the forecasts of ever higher world demand for energy. This led to the conclusion that Britain would be left on the sidelines if it was unable to meet its own growing energy requirements without resorting to imports for which it could be held to ransom. Indeed, energy consumption forecasts for the year 2000, which were disclosed in the 1987 Energy Committee report, still showed a potentially massive demand for primary energy in the UK by 2000. These forecasts appear to have been drawn assuming a straight line projection of future growth from the year in which they were made and tend not to take account of the cyclical upswings and downswings which are bound to occur in any economy. As such the forecasts for energy consumption gradually declined over the period as the actual primary energy demand in the country also declined due both to the recession of the early 1980s and the increasing energy awareness of industry as a whole (see Table 11.2). Indeed, this energy awareness is something which is likely to continue in the future despite the relative cheapness of primary energy in the early 1990s in comparison with the early 1980s. It will also be encouraged if the British government, and others overseas, bring in a carbon tax in order to attempt to limit the amount of both carbon dioxide and other potential greenhouse gases produced and emitted into the atmosphere. The growth in developed country energy consumption is therefore expected to take place at a lower rate than gross national product (GNP).

In 1991 the UK's primary energy consumption was about 329 million tonnes of coal equivalent (mtce), slightly higher than the 321 mtce and 318 mtce of 1989 and 1990 respectively, but still less than the 356 mtce consumed in 1979. That energy consumption is rising despite a recession is somewhat surprising, although the declining cost of energy due to the fall in the price of oil must play a significant role in this. The 1979 to 1982 fall

Table 11.2 Various forecasts of energy demand, 1976–82

Year of forecast	Author of forecast	Actual demand in year of forecast (mtce)	Forecast primary demand in 2000 (mtce)
1976	Workshop on Alternative Energy Strategies	330	490–610
1976	Department of Energy	330	420–760
1977	Department of Energy	338	500–650
1978	Department of Energy	340	450–650
1978	JH Chesshire and AJ Surrey, Sussex University	340	335–577
1979	G Leach International Institute for the Environment & Development	356	330–361
1979	Department of Energy	356	445–510
1979	C Robinson	356	403–450
1980	World Coal Study	329	360–450
1982	Department of Energy	312	328–461
1982	Central Electricity Generating Board	312	256–418

mtce = million tonnes coal equivalent.
Source: House of Commons Energy Committee Report on the Coal Industry, 1986/7 session, HMSO.

in UK energy consumption occurred at a time of high energy prices, which in part caused the recession in the first place, and so companies included energy conservation strategies as part of their cost containment programmes. Such strategies are lacking from the current recession which has been caused much more by financial than by energy supply/demand factors.

British Coal's further problems during much of the post-World War II period were exacerbated by the changing structure of the energy market within the country. In particular most of the demand for coal in post-war Britain came from industry as a whole, with specific demand from iron and steel, domestic demand, commercial and public services and transport. Indeed, these areas accounted for up to 80% of total coal consumption during the period with only some 20% of coal being consumed in power stations for electricity generation. This situation has now been completely reversed with the loss of many of these markets for British Coal and the use of some 80% of total coal output in the power station as opposed to the industrial market. Finally, the electricity market also started to move away from coal following the development of nuclear power stations after 1956 and although nuclear energy only accounted for some 7.2% of total primary energy demand in the UK in 1991 it still represented some 20% of total electricity supply.

Since 1983 the pace of the attrition in the coal industry has speeded up. The loss of 140 pits between 1983 and 1992, combined with the reduction in the workforce from an average of 207 600 in 1983 to only 51 000 before the closure announcement in 1992, shows the scale of the cutbacks which have taken place. This is against the background of the government's obvious desire to restructure the industry in order to prepare it for privatization and Mrs Thatcher's desire for this to happen sooner rather than later. The appointment of Neil Clarke to manage the industry in the run-up to privatization was one of the main planks of the programme. He has stated that he did not want to close so many pits and throw so many miners out of work. However, he has no long term desire to stay in the mining industry in Britain and, as soon as the privatization has taken place, would prefer to go off and grow roses than stay and look after the industry which he helped to create. This apparent lack of commitment cannot be good for morale within the industry and must be behind some of the press speculation that his tenure as Chairman of British Coal will be foreshortened.

CAPITAL CRAVINGS

British Coal's major problem within the public sector has been capital. Because the company has not been able to raise any equity capital it has been forced to rely on the government for finance. This money has been raised both through a deficit grant and through borrowings. The deficit grant has been used to fund much of the company's redundancy and other programmes necessary to reduce the number of pits and the size of the workforce. The size of the grant has depended on the amount of money which British Coal has needed in any one year to meet these obligations and has therefore fluctuated widely depending on the scale of the job cuts. British Coal's borrowings from the government have a long history and go back to the structure of the industry at the time of nationalization in 1947. At times the level of this debt has been high and British Coal has consequently had to pay high amounts of interest to the government. That this is just money changing hands from one pocket to another within the Treasury should not be forgotten – especially because the company would have had access to equity capital in the private sector and would not have needed to pay such large amounts of interest.

In its report in 1987 the Energy Select Committee covered this point in a footnote where it stated that 'nationalized industries are predominantly debt-financed and so have to pay interest on their loans from the Treasury for investment, irrespective of their short-term trading performance'. It mentioned that 'a welcome step . . . would be for more of the industry's capital requirements to be funded on an equity rather than a loan basis, so that the 'shareholders' [in effect the government] would reap high returns in good years and low returns in bad years'. That British Coal was not able to do this has meant that the company has been starved of capital and has been unable to proceed with the investment necessary for its future as soon as it might otherwise have liked.

Additionally, the apportionment of the interest on this debt must be one of the reasons why British Coal is able to insist that some pits are losing money despite the local management protesting otherwise (note the discrepancies at Grimethorpe highlighted by the Boyd report into the ten pits on the definite closure list – see Chapter 17). This appears to be the new law of diminishing returns whereby central and overhead costs are apportioned equally across a number of different divisions. If one division is losing money it is closed down, even if it is making a profit within its own operating environment and even after taking account of capital and other financing needs. The remaining operating units then have the same level of overheads split between fewer members and as these attributable costs have risen, then another division is seen to be losing money. The end of the process occurs when the company becomes a central cost with no operating units actually generating any profits. This situation has occurred many times across all sectors of business and finance and often occurs when the 'bean-counters' gain control.

It must, however, be admitted that the total closure of British Coal is unlikely despite the appearance that this is inevitable. Indeed, even many of the mines in the closure list seem to be making money at the operating level and may be able to reduce operating costs to such a level as to remain profitable in the lower coal price environment expected in the UK in the future. The most obvious examples of this are the inclusion of both Grimethorpe and Betws Drift in the list of 10 mines which are to be closed after the statutory 90 day consultation period. Neither of these mines services the power station steam coal market, with Betws Drift producing speciality anthracite and Grimethorpe speciality coking coal for ICI's nearby Monckton plant. ICI was given no notice of the impending closure of Grimethorpe and is now being forced to look elsewhere for its supplies of coal and may be forced to resort to imports if no local supply is available.

THE FINAL COUNTDOWN

The closing of the Thatcher period left British Coal in a much weaker position than it was in 1979. The company now only has some 31 pits which will remain in operation and is likely to move back into loss in 1992/3 and 1993/4, as a result of the redundancy packages and the lowering of the price of coal to the electricity generators, respectively. Other British Coal operations, which are not specifically in the coal mining industry, have been sold off during the past 13 years. These sales have been part of the reason why British Coal has been able to reduce its debt. However, it means that the industry is now more fragmented than in the past and that it is, as a whole, less secure because of its increasing reliance on the electricity generators. This means that the government will raise less money from a sale of the coal industry in its constituent parts than could have been achieved through the sale of some larger, integrated units.

The Thatcher legacy has, therefore, been disastrous for the British coal industry as it has been based on the precept that the free market will only work if there is no vertical integration between producer, supplier and consumer. This is the main problem facing British Coal today because it no longer has any control over its end markets. The central rump of the industry has been left in a completely uncompetitive position with regard to imports and the company is no longer protected by its downstream industrial activities. The one exception is the royalties obtained from open cast licensing, although these too are likely to cease at the time of privatization. There has been no stimulation of competition by the split of the company back into area boards with their own distinct operations, financing needs and profit centres and thus the highly restrictive practice of vesting all of Britain's coal in the Corporation has continued. Whilst it may be true of the unions it seems difficult to see why the management would have an interest in preventing any other operator from constructing a new mine, whether underground or open cast, except that this may show their own inefficiencies within the state run enterprise.

On top of this the morale within the Corporation is reported to be at such a low ebb that many employees are either handing in their notice and obtaining massive voluntary redundancy payments, or are waiting either for retirement or compulsory redundancy. Such a low level of morale in the head office may be partly behind the delays experienced by the DTI in preparing its White Paper on the future of the industry. It is also the case that the Trade and Industry Select Committee, among other investigators, has complained about the length of time that it takes to get an answer out of British Coal. And whilst interest is again surfacing new investors in a privatized industry will still have this major hurdle to overcome.

Chapter 12

The electricity conundrum

THE LAST FRONTIERS

The electricity generating and metallurgical industries are the two largest single users of coal in the
world today. Of these the use in the electricity industry is the largest and most
important. The rise in the use of coal in electricity generation in Britain can
be seen from Figure 12.1. This displays the amount of coal used in tonnes and
the percentage of the total consumption of coal it represents. Figure 12.2
shows the split of the electricity generation market in the UK, and shows how
the dominance of coal has been eroded over the years. Such a situation is not
peculiar to Britain as can be seen from the splits shown for other countries in
Figure 12.3. Even China, with its great coal reserves and production, has
been diversifying away from coal as more hydro-electric plants are either
under construction or are being designed.

The Clean Air Act which was introduced in 1956 initially encouraged the
demand for coal in the electricity industry as it led to a much higher demand
for electricity for home heating. In effect the pollution generated by burning
coal was removed from town and city centres to areas in the countryside where
the carbon dioxide was less concentrated and could be dissipated by the wind.
The introduction of smokeless fuels by British Coal helped to regain some of
the domestic market but the loss of many primary customers meant that British
Coal became increasingly dependent upon the electricity industry for the main

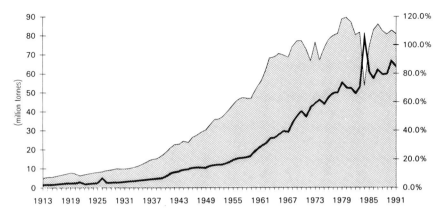

12.1 Coal use in electricity generation and share of total demand 1913–91

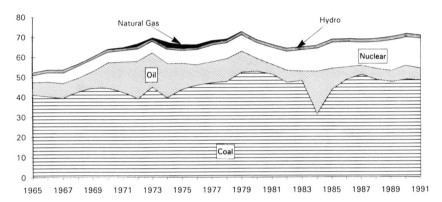

12.2 Energy source for electricity generation 1965–91 (million tonnes of oil equivalent)

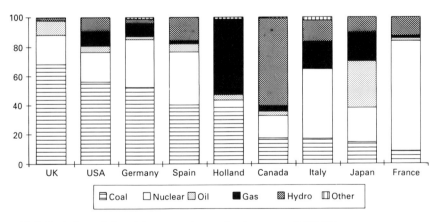

12.3 Fuel share of electricity generation market by country 1990 (%)
Source: HMSO UK Energy Statistics, OECD Coal 1992.

159

part of its output. And in the industrial sector coal's share of total demand was eroded both by the need to keep within the requirements of the law and by the increasing level of competition from other energy sources. As an example in 1956 ICI calculated that it was cheaper for the company to run its ammonia plants on oil rather than coal and, subsequently, in 1958, the change-over was completed.

The move away from coal, both in direct consumption and indirectly through the increasing diversification of energy sources in the electricity generation industry, lies at the heart of the electricity conundrum. For, as greater knowledge of other energy sources has been obtained, and their relative cost ascertained, it appears only to have been legislation that kept the majority of the coal industry's mines operational. This legislation has, in part, been forced on successive governments because of the acknowledged power of the mining unions to bring the economy to a standstill. Until now it also prevented the need for a truly economic appraisal of the cost of coal-fired electricity. The major constraint on such an appraisal was the restriction on the import of coal until 1979 as it removed the need for British Coal to improve productivity. This meant that it did not have the full incentive of unfettered competition and that, even following 1979, productivity did not improve at the rate needed to maintain market share when all the barriers to entry were withdrawn. This has still to occur as the renegotiation of the 1990/3 coal supply contracts and the 1993 White Paper continue to protect the higher cost areas of the coal industry from the lower cost competition overseas.

It was during the renegotiation of the coal supply contracts during 1992 and early 1993 that the complicated nature of the conundrum first came to light as it was only then that the potential erosion of coal's share of the electricity generation industry raised questions in the public's mind. Indeed, the conundrum is not just about which fuel is cheaper in the conventional sense of a cost per unit of energy, it is about how effectively the fuel can be used. This is related both to the amount of energy that can be converted into electricity and the cost of switching power stations on and off to take account of peaks and troughs in demand.

A reduction in the use of British coal by the electricity generators has also caused a build up of stocks at power stations and at the pit-head. In late 1992 there was a total of some 46 mt of coal stocks, 30 mt of which was at the power stations and 14 mt at British Coal. The remaining 2 mt was held in industry stockpiles. The historical level of coal stocks in Britain is shown in Figure 12.4 which clearly exhibits how strong the bargaining position of the two power generators was during the negotiations. This is because there were sufficient stocks of coal in the country, already dug out of the ground, to meet

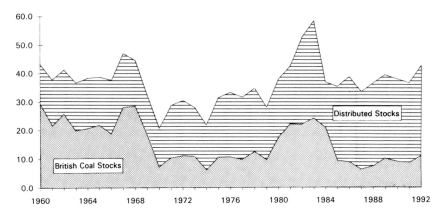

12.4 Total British Coal stocks 1960–92 (million tonnes)
Note: 1991/2 figures are for end March.

the planned demand for coal for the whole of the 1993/4 financial year. As a result British Coal was forced into a very weak bargaining position and because it could not find a solution which would have kept all of its mines in operation it was forced into the October 1992 announcement to close 31 pits because, as Neil Clarke said, 'there is not a market for our coal'. And the gift of a government subsidy to keep 12 extra pits in operation between 1993 and 1995 still fails to address the conundrum which, it is now hoped, will be solved by the wherewithal of the productivity conscious free market.

THE GAS DILEMMA

The further problem facing British Coal in its negotiations with the electricity generating industry during most of 1992 was the capacity of the gas-fired power stations which were either being planned or built at the time. In some respects it seems that these stations were also being built with government blessing as the lower amount of pollution they produce was widely held up to show that the country would be able to meet the aim of restricting the amount of carbon dioxide produced in 2000 to the amount produced in 1990. The burning of gas is about 50% effective, as opposed to the effectiveness of around 40–48% for coal-fired power stations. This makes gas a better fuel to use in this application than coal – especially because some of the heat is generated through the burning of the hydrogen in the CH_4 molecule to produce water. This means that there is a lower output of carbon dioxide for each unit of heat produced and gas is therefore a cleaner fuel to use in power stations.

However, the problem for natural gas is that the specific components of the molecule make it a very useful and important building block for the chemical industry in addition to its basic heat production possibilities. The fear that the use of gas in areas which could not be designated as premium markets (i.e. ones which could effectively be shown to be wasteful of a scarce resource) would lead to an early run-down of British reserves thereby increasing the need to import gas from overseas or to invest heavily in the production of synthetic natural gas (SNG) is not a new angle to the energy debate. It was this fear which prompted Sir Denis Rooke, the former Chairman of British Gas, to promote the need for a concerted energy policy in the UK, the essence of which 'should be to encourage the correct use of the different fuels in the economy'.

Proven gas reserves which can easily be utilized in the UK are sufficient to last for some six years at 1992 rates of consumption but more deposits are known so that the industry has confidence that there are reserves sufficient for at least some 20 years. With the increase in consumption that is anticipated if all of the proposed gas-fired power stations are constructed, these reserves are not likely to last so long. In his statement to the House of Commons in October 1992 Michael Heseltine estimated that the country's gas reserves were sufficient for about 55 years. This included the country's gas resources and made an allowance for reserves and resources which were not yet discovered and which it is hoped and expected will be discovered in the future.

What he did not say was that there would probably have to be a considerable rise in the gas price to make some of the known resources economic and that, whilst this would encourage exploration to find further reserves and resources, it would also lead to much higher prices for the consumer. Indeed, a US survey has estimated that the international prices of oil and gas will increase eight times faster than the price of coal during the first 10 years of the next century. This clearly shows the disadvantage that coal is faced with over the short term when compared with its longer term prospects.

The gas dilemma is, however, further confused by the efficiency of using gas for home consumption compared with its use in the generation of electricity. This is because when gas is used for either heating or for cooking its efficiency approaches 100%. This means that only half the amount of gas would need to be burnt to bring a pan to the boil on a gas cooker than would need to be burnt to generate the necessary electricity on an electric cooker. It also does not take into account the energy loss which results from the transmission of electricity over long distances. Aluminium, copper and steel are all resistors and generate heat when electricity passes along them in power transmission lines. This means that there is a continual energy wastage which further reduces the efficiency of electricity against the other fuels. This is taken into account by the National Grid Company because of the greater weighting of power stations in the north of England. As a result consumers in the south of the UK are charged a premium price for the electricity they use and large industrial consumers such as English China Clays, which anticipates a 10% increase in annual costs (equivalent to £500 000/year), are put at a considerable disadvantage. Indeed, the consumption of coal in an Aga or other combined cooking/heating unit would probably be more energy efficient than using electricity.

Until 1990 the restriction on the use of gas was a relatively easy task as the government was prevented under EC guidelines from allowing the use of natural gas in electricity generation. EC energy policy was therefore dictating the energy policy of the UK and, at the same time, protecting the electricity generators in the UK from the onslaught of the Regional Electricity Companies (RECs) who were becoming increasingly concerned about the fairness of the pool purchasing market for electricity. Indeed, they believed strongly that the two big generators, National Power and PowerGen, were effectively controlling the market and were not passing on the full benefits of either lower coal prices or better operating efficiencies to their customers in the shape of lower electricity prices. As stated in the previous chapter, this is one of the legacies of the Thatcher administration which decided to push ahead with the privatiz-ation of the two big generators in March 1991 despite the lack of a need to do this when the nuclear electricity industry was taken out of National Power. In order to create a fairer and more competitive market the government could have, and indeed should have, split the generators up into smaller units and many of the RECs believe that a total of five generators could have been created, thereby increasing the competition in the market substantially.

However, the lack of competition encouraged the RECs to look into the possibility of constructing their own power stations in order to diversify their source of supply away from the two big public generators and Nuclear Electric. The first plant, at Roosecote in Cumbria, was approved by the Secretary of State for Energy in 1989 (i.e. before he was strictly allowed to do so under EC rules) and was commissioned in October 1991. The plant is owned by Lakeland Power, a joint venture company in which Norweb holds a 20% stake. It has a rated capacity of 220 MW of electricity but only supplies around 8% of the electricity consumed in the Norweb area. Norweb is also a partner in a joint venture to build a second plant at Keadby, on Humberside, which has a much larger capacity of 750 MW, although it will only be buying around a third of the output. The likely dates of the plants which are expected

Table 12.1 Planned gas-fired power stations and their capacity

Name	Location	Capacity (MW)	Start date
Operational:			
Roosecote	Cumbria	220	Oct 1991
Killingholme	Humberside	900	1992
Wilton	Cleveland	1875	March 1993
Under construction:			
Sellafield	Cumbria	100	1992/3
Brigg	Humberside	272	1993
Corby	Northamptonshire	412	1993
Killingholme	Humberside	650	1993
Peterborough	Cambridgeshire	348	1993
Barking	Essex	1000	1994
Deeside	Clwyd	450	1994
Keadby	Humberside	750	1994
Rye House	Hertfordshire	790	1994
Medway	Kent	660	1995
Spondon	Derbyshire	318	1995
Total capacity		8745	

Sources: ENRON, TUC, National Grid, Coalfield Communities Campaign.

to be brought on stream in England and Wales over the next three years, together with their capacity, are shown in Table 12.1.

At one stage the National Grid Company knew of a total of 22 876 MW of capacity that was either planned, under construction or in operation. Its latest figures show that some of this capacity has now either been put on hold or completely cancelled. In particular it stated on 23 November 1992 that 2213 MW of potential supply had been cancelled and that a further 3771 MW had been deferred. Nevertheless, this still leaves a total of some 6891 MW which is still likely to be added to the list in Table 12.1, making a total of some 15 646 MW of gas-fired power station capacity which will probably be in production by the end of the century. This is likely to displace some 38 mt of coal a year from the electricity generation market.

THE DASH FOR GAS

It was the success of Norweb's first plant at Roosecote which initiated the so-called dash for gas. None of the other RECs wanted to be left on the sidelines with the requirement to purchase electricity from the two big generators without the ability to negotiate. Roosecote provided the proof that under the British Gas and other gas purchase contracts they were being offered they could produce electricity at a lower price than they would have to pay to the generating companies. This was because of the excess gas which British Gas was able to supply at the time and the requirement by the gas regulator, Ofgas, that British Gas publish a listing of the prices and conditions which it would attach to long term gas purchase contracts. This first led to the publication of LTI2 (LTI stands for long term interruptible) in November 1990. As demand for gas increased in the electricity supply market a second schedule of prices, LTI3, was introduced on 17 September 1991, and with still strong demand a third price rise was instituted on 15 October 1991.

The supply of gas on long term interruptible contracts was deemed necessary by the RECs because of the need to ensure long term availability of gas to the power stations at prices which the companies deemed would enable them to ensure that they were able to provide electricity cheaply and efficiently to their customers. That the contracts were interruptible means that the supply of gas can be switched off (after a given period of notice) for a specific number of days as detailed in the contract. This is necessary for the gas supplier because of the increase in demand for gas which normally occurs during periods of bad weather. The pumping of gas out of the wells cannot be increased greatly to meet such periods of increased demand and so the gas suppliers attempt to ensure that they can reduce supplies to certain large industrial consumers in order to meet the needs of their retail customers. This is particularly important because of the dangers of an explosion which may occur if there is a drop in the pressure in the gas supply system.

In order to ensure that they achieve some of the increase in the ruling price of energy in the market the gas producers put an escalation clause into the price of the gas contracts they offer. In the LTI2 and LTI3 contracts British Gas allowed the purchasers the choice of one of three different types of indexation in the contracts, which related the future price of the gas supplied more or less heavily to the price of oil, or to the producer price index and either electricity or coal. Different weightings meant different starting prices for the contracts and because of the relationship which generally exists between oil and gas prices the contracts with a higher oil price component in the indexation factor started from a lower base. The use of an oil-based escalation factor allows for the fluctuations in the sterling exchange rate as a higher world oil price in US dollars will feed through into sterling at the ruling exchange rate. If sterling is relatively weak then this will lead to a further increase in the sterling oil price and thereby increase the indexation factor for

Table 12.2 The three types of gas contract offered by British Gas

Escalation type:	A	B	C
	15% Gas oil	20% Gas oil	25% Gas oil
	15% Heavy fuel oil	20% Heavy fuel oil	25% Heavy fuel oil
	35% PPI	30% PPI	25% PPI
	35% Electricity or coal	30% Electricity or coal	25% Electricity or coal
Contract	Pence/therm	Pence/therm	Pence/therm
LTI2	17.00	16.50	16.10
LTI3 (Old)	20.30	19.80	19.60
LTI3 (New)	21.20	20.70	20.50
Metric version	Pence/kWh	Pence/kWh	Pence/kWh
LTI3 (New)	0.723	0.706	0.699

Source: British Gas.

the gas contracts. The escalation factors and prices of the three contracts are detailed in Table 12.2.

In its evidence to the Trade and Industry Committee inquiry in late 1992 British Gas stated that it increased the price of its contract so rapidly between September and October 1991 because the massive demand from the RECs and their joint venture partners continued. The company therefore needed to attempt to choke off this demand because of the long lead time necessary to build the infrastructure required to ensure that the demand could be met. The second rise in the price was sufficient to stifle this demand but the company has said that if there is any significant interest shown at these levels then it will raise its price again as it would not be in a position to meet any major interest for a substantial period of time. This policy is still very much the line taken by Sir Denis Rooke, who said that 'the desires of the customers naturally do not always neatly fit the pattern of gas availability and the Industry has therefore had to adopt various measures to reconcile supply and demand within economic constraints'.

It can therefore be assumed that whereas the electricity producers thought that they could generate electricity economically at prices in the high teens in terms of pence/therm (the implication is that they opted for escalation types B or C as opposed to escalation type A) it would not be economic, or at least provide a satisfactory return, at a price of just over 20p/therm. The price unfortunately gives little indication of whether or not the electricity would be produced within the regulations laid down by the Electricity Act, and enforced by Offer. It is, however, fair to assume that the margins offered would still be within the prices offered by the two big generators, even at the higher price.

The rise in the price of the gas offered by British Gas has not totally stopped the building of new gas-fired power stations because of the open playing field in gas supply which was created at the time of the privatization of the gas industry. This means that, although the major supplier, British Gas is not now the only supplier of gas to the electricity generation market. The country's gas reserves will therefore be used at a faster rate than would have been the case had British Gas remained the sole supplier to the UK market. In spite of this there must be a benefit to British industry if the lower gas prices are passed on to the consumer. It is this price and profit question which lies at the centre of the electricity conundrum.

THE NUCLEAR QUESTION

The other main area of confusion in the electricity debate surrounds the future of the nuclear generation industry. This is because the costs of nuclear power are so confused and confusing that it is almost impossible to divorce fact and fantasy. The whole argument about the existence of the nuclear power programme relies on the cost of the decommissioning of the stations and whether or not this, and other, costs fall into the avoidable or the unavoidable area.

The reason for this line of argument is that the future use of coal could be increased if the seven Magnox power stations which are currently being used to generate electricity by Nuclear Electric were closed down over the next few years rather than later in the decade and into the early years of the next century. Early closure of the seven stations would remove some 3334 MW of capacity from the electricity network, equivalent to some 8 mt of coal. This would obviously be of considerable benefit to the coal mining industry, but the question of the closure of the nuclear power stations is one for the government rather than for the market. Strangely enough the market was never supposed to affect 'the UK's first full scale programme of commercial nuclear power [as it] was based on the Magnox reactors which were not expected to compete on cost with contemporary fossil fuelled plant' according to Sir John Hill, a past chairman of the nuclear industry.

Indeed, it should not be forgotten that the government provided for the decommissioning of the Magnox stations during the time when they were in operation under the aegis of the old CEGB. However, owing to the inadequacies of the old CEGB's accounting practices the amount provided was insufficient to cover the cost of decommissioning, and electricity consumers are now being asked to make a much larger contribution to the fund. It should further be remembered that the quest for nuclear energy was one of the government's own instigation to ensure that Britain was a world force in the nuclear and defence industries. Indeed, the first Magnox power station opened was that at Calder Hall which remains under the control of British Nuclear Fuels Limited (BNFL) rather than having been transferred to Nuclear Electric at the time of the creation of the latter company. This station, and its sister station at Chappelcross, were designed to produce plutonium in order to aid the Ministry of Defence rather than to generate electricity.

The early closure of these stations, whilst it would be beneficial for the coal mining industry, would not be to the advantage of either Nuclear Electric or BNFL. As far as Nuclear Electric is concerned it is due to receive the nuclear levy to cover the costs of decommissioning the seven stations between now and 1998. After this date the company should have raised sufficient funds, together with the cash flow generated from running the plants over the intervening period, to start on the decommissioning process. This process necessarily takes a long time and costs a large amount of money and the company needs to be certain that it will be able to cover its liabilities, which could extend for up to 1000 years, before it can be certain that the decommissioning process should be started.

One of the unsettling revelations to come out of the Trade and Industry Committee inquiry into the future of the British coal industry is that the nuclear industry is, in fact, spending some of the cash generated from the levy to build the Sizewell B power station. The money is therefore not being put aside into a separate pool for the closure programme, but is rather being spent on another programme which will, in turn, need to generate income in order that it can be decommissioned successfully. Obviously, Nuclear Electric expects that Sizewell B will generate a large amount of free cash when it becomes operational in 1994 but the excessive £750 million cost of the plant probably means that it may never provide a positive return on the initial investment.

THE PRICE OF POWER

Despite the efficiency arguments, much depends on price. The use of energy equivalent costs has now moved into the electricity generation market where the cost of production of electricity has been com-

Table 12.3 Various estimates of the cost of electricity generation, comparing the different energy sources (pence/kWh)

Date	Source	British Coal (old)	Old coal with FGD	New coal	Imported coal	Gas	Nuclear	Oil
1979	The Energy Debate	1.56	—	—	—	—	1.30[1]	1.93
1991	PowerGen	2.73	—	—	2.19[2]	2.64	—	—
1992	British Gas	2.2–3.6	2.7–4.1	3.6–4.2	1.6–2.9	2.3–2.7	8.2	4.2
1992	Coalfield Communities Campaign	2.1–2.7	—	—	—	2.7–3.3	6–7[3]	—
1992	Electricity Supply TUC	2.6	—	—	—	2.7–3.0	—	—
1992	Enron Europe	2.53–2.68[4]	3.19–3.34	3.5	1.60[4]	2.65	—	—
1992	Major Energy Users Council	—	—	—	1.50	—	—	—
1992	McCloskey Coal Information	2.4	—	—	—	2.4–3.2	4.0–8.5	—
1992	National Power	2.60–2.75[4]	3.2–3.35	—	2.05–2.2[4]	2.4–3.3	—	—
1992	Nuclear Electric	—	2.8	—	—	—	2.8–5.0	—
1992	PowerGen	2.2	—	—	1.35[4]	—	—	—
1992	Scottish Nuclear	—	—	—	—	—	3.2	—
1992	Southern Electric	—	—	—	—	2.65–2.8	—	—
1992	Sunday Times	3.0	—	—	—	2.5–3.2	5.2	—
1992	Sussex University	—	—	3.5–4.0	—	3.0	7–8[5]	—
1992	TUC	1.9–2.9	2.42–3.42	—	—	2.0–3.3	—	—
1992	UDM	2.5	—	—	—	3.3	—	—
1993	Eastern Electricity	3.26[6]	—	—	—	—	—	—
1993	National Power	—	2.9–3.05	—	—	—	—	—
1993	Scottish Nuclear	—	—	—	—	—	3.0	—
1995	Scottish Nuclear	—	—	—	—	—	2.5	—
1996	Sunday Times	2.2	—	—	—	—	—	—
1998	National Power	—	2.65–2.80	—	—	—	—	—

[1] Includes provision for decommissioning.
[2] Delivered to inland station, includes 0.53p/kWh FGD capital cost – all other imported coal assumed to be low sulphur which does not need FGD equipment to be installed.
[3] French nuclear generation costs estimated at 2.5p/kWh.
[4] Based on source's figures.
[5] Existing reactors (Sizewell B estimated at 6p/kWh – elsewhere estimated at 7.3–9.2p/kWh).
[6] Estimated contracted purchase price.

pared endlessly in the preparation of many reports and studies into the future of the coal industry. These comparisons have been many and varied and some of them are represented in Table 12.3. The difficulty in comparison is simply that the final number, which is supposed to show the relative cheapness or expensiveness of the various methods of electricity generation, is not prepared on a comparable but rather on a current cost basis. This means that in one respect like is not being compared with like and it appears that this may be the intention of some of the power generators when they are attempting to prove to their regulator that they are supplying the customer with electricity as cheaply as possible. If they are stating that the electricity is produced as cheaply as possible assuming that a new power station had to be built then perhaps they would be correct. The problem is that no new power station has to be built as there are plenty in operation which would be forced out of the market if new stations were built. This is even though the cost of producing electricity from these stations is lower than the cost of electricity produced from a new gas station as all of the capital cost of the old coal-fired power stations has been written off and there is only a minimal depreciation charge to take account of in the cost calculation.

The other problem with respect to cost covers the use of power stations as suppliers of base or peak load electricity. This distinction is important because the efficiencies of all power stations increase with increasing use as more of the fuel burned is converted into electricity rather than being lost in cooling during periods of inactivity. All of the nuclear stations (with a capacity of some 8300 MW or nearly 22 million tonnes of coal equivalent) run on base load, as will many of the gas-fired stations. This will leave the coal, hydro-electric and other plants to run at peak times and as a result they will have a lower efficiency and higher costs. This is one reason behind the closures of the older and least efficient coal-fired generating plants announced by both National Power and PowerGen in early 1993.

It is also apparent that different interest rates are being used in the calculations. In one instance a cost of capital of only some 5% was used and this showed that the cost of a new gas station was economic. It was not clear where the company concerned would be able to borrow money at such a low rate of interest, when market rates are considerably higher, despite the government's new policy for economic growth. The main reason for this need to justify the building of new electricity generators is the desire of the electricity distribution companies to have some control over their supply rather than being forced to buy it from an effective oligopoly where what the much larger National Power says goes.

The playing field is therefore highly pitted and piled and it is no wonder that the electricity regulator, Professor Stephen Littlechild, was originally intending to take about a year over his investigation into the pricing structure within the industry. However, the timetable was speeded up following the public furore after the pit closure decision and the results were released in early 1993. They confirmed that the cost of the new gas generators was not excessive and that the RECs were acting within their obligations to provide electricity cheaply. As a result the construction of gas power stations was allowed to continue and British Coal was restricted to fighting for the smaller section of the electricity generation market left for coal.

Chapter

13

Long term demand

THE SECOND COAL AGE

The world's reserves of oil and gas are now expected to be sufficient to maintain the current rate of output for up to 55 years. After this time the depletion in the reserves and the increasing cost of extraction, as oil companies are forced into yet more hazardous and inhospitable environments, will start to reduce the competitiveness of oil and gas as sources of energy. Admittedly these scare stories have emerged before, and despite the impending change of century, the world has not run out of oil as the forecasters of the 1970s would have had us believe. It is, however, true to say that all of the carbon-based energy sources on the planet are limited and that the increasing rates of economic growth around the world will lead to greater rather than reduced demand for energy. This implies that reserves of oil and gas will be exhausted sooner rather than later.

Obviously the use of oil and gas is related to two factors. One is ease of use and supply and the other is price. It is cheaper to run a car on petrol than it is to run it on coal which has been broken down and reconstituted along the lines forced on South Africa during the oil embargoed years of sanctions. The depletion of the rail network is further evidence of the cheapness of road transport – especially because of its ability to move goods directly to where they are needed. In the days of rail most factories of any significant size had their own railway spur which reduced the need for road transport. The exceptionally high rates of inflation suffered during the post-World War II years encouraged companies to reduce their working capital requirements as these funds could better be invested in capital projects. This, in turn, led to the implementation of just-in-time delivery systems which meant that companies could not leave goods to sit in railway marshalling yards for extended periods of time. Road haulage was, and is still, used instead.

As the cost of petrol and diesel fuel increases as a result of the exhaustion of reserves all trading companies will have to look for alternative methods of transport and power generation. One of these is likely to be a move back to the railway network. This would probably be powered by electricity generated in distant power stations rather than directly by coal following the re-emergence of the steam engine. And, in view of the world's dependence on the internal combustion engine there is likely to be a switch to making oil from coal on a major scale as soon as it makes economic sense to invest in the necessary technology. Notwithstanding the invention of an alternative method of transportation, and the advent of the electrically powered car (which itself is likely to be recharged from coal-fired generators), the need for petrol or diesel engined vehicles is likely to be maintained − especially in areas with an inefficient electricity distribution network.

One final concern surrounds the environmental destruction that is associated with any use of fossil fuels. Indeed, this tended to put both Friends of the Earth and Greenpeace campaigners into something of a quandary during the debates on the future of electricity generation in the UK. In the first instance they do not approve of the use of nuclear power stations because of the slow leakage of radioactive waste and, more importantly, the potentially devastating effects of an accident. However, the cumulative effects of acid rain and carbon dioxide emissions also mean that they cannot directly advocate the use of coal − and even less English and Welsh coal in view of its high concentrations of sulphur.

Finally, as mentioned in Chapter 12, the relatively inefficient use of gas in gas-fired power stations is a cause of concern for the environmentalists, although the new Combined Heat and Power (CHP) schemes, whether coal or gas, are preferred because they are the most efficient. As even the apparently benign use of hydro-electric power causes problems due to the artificial alteration of the water table and loss of the natural landscape, a move to greater emphasis on conservation rather than generation remains the main thrust of their campaigns. This aim was helped by the Chancellor of the Exchequer when he announced the future imposition of VAT on domestic fuels in his March 1993 Budget.

Whilst admitting that necessity is the mother of invention and that other, cleaner, sources of energy are likely to be found in the future, the rapid depletion of the world's oil and gas resources seems to give little thought to this impending denouement. Therefore, as there currently seems little alternative, it must be expected that coal will stage a major recovery in demand from the second quarter of the next century. This will be the second coal age.

THE ENERGY QUESTION

One of the main factors in determining the long term demand for coal within the UK is the need to ascertain what the country's energy demand will be in the future. From this figure a distribution can be constructed to estimate how much of which fuel will be consumed to meet the total demand and then this demand pattern can be used to predict the breakdown of suppliers. In determining the fuel supply distribution, other outside factors need to be taken into account, including the environmental and balance of payment impacts of any decision.

In Chapter 10 an indication of the current and projected supply/demand balance for UK coal was given but it was necessarily brief. The reasons behind the variance in the projections for any supply/demand balance have also been covered earlier and whilst it is a difficult task it is necessary to attempt to forecast future energy demands within the UK because this provides the foundation on which any future use of coal will depend.

Energy demand in terms of million tonnes of oil equivalent from 1950 to 1991 is shown in Figure 13.1. Real GDP and the energy consumption/GDP ratio are also depicted. This latter figure is important because it shows the energy consumption needed by the UK to produce a specific amount of GDP. If it is projected that GDP will continue to rise, over time, then it can be assumed that the country's energy consumption will rise by a similar amount. This figure will, however, be influenced by environmental aspects which lead to a higher energy efficiency target in the future compared with that which exists today and that which was in place in advance of the first oil shock of the early 1970s.

The future will also be distorted by such items as the country's economic cycle and so the use of energy will decline, or will at least not grow so quickly, during periods of recession and depression. Finally the country's energy consumption will be related to the weather. In any one year the average temperature should influence energy consumption because of the large

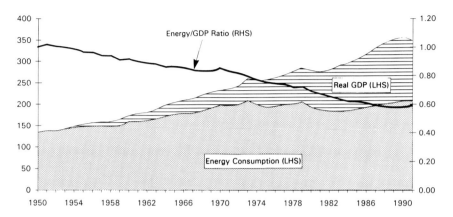

13.1 UK real GDP, energy consumption and energy/GDP ratio 1950–91
Source: Department of Energy.

domestic energy usage in home heating. This is allowed for in Figure 13.1 because demand has been temperature corrected. It is, of course, not possible to predict the weather even further than a week in advance with any accuracy, whilst the effects of global warming cannot yet be detected with any real certainty and are unlikely to bear any real influence over the period of this forecast. In some respects the energy savings which should come from a higher mean temperature can be included within a factor to account for increased energy awareness within the industrial community and, therefore, should be included in overall energy savings figures. Also the UK does not have the hot summers of the USA which boost energy consumption through increased use of air conditioning units. Indeed, British energy use during the summer months is significantly below that in winter and although this will change with the increased use of CCGT stations the winter/summer gas consumption ratio is of the order of 6:1.

The economic cycle of the UK is, additionally, increasingly difficult to forecast, because of the increasing influence of the policy decisions taken not just elsewhere in Europe but elsewhere in the world. In the nineteenth century, when Great Britain was the largest energy user in the world, the country dominated world politics and economics and policy decisions were more understandable in this context. In the final quarter of the century the rate of expansion elsewhere in Europe and in the USA overtook the rate of growth in Great Britain and the country's influence on world events started to diminish. It was during this period that Britain went through a major period of deflation and so-called depression despite the fact that GDP continued to grow.

In this case many of the policy decisions taken by indigenous governments, of whatever political persuasion, may have little overall effect on UK energy consumption in the future. As the country becomes more and more inextricably linked to the vicissitudes of the European market, and that market continues to expand with the integration of yet more countries, the UK's influence on its own economic progress is likely to be eroded. This will be especially the case following the introduction of a pan-European currency because such an event will remove a political tool which has been used time and time again in order to attempt to restart growth within the economy. This loss of political influence over the direction of an economy may well be beneficial over the long run as it will remove the possibility that a country can 'go for growth' and thereby rekindle all of the inflationary ills which brought about the period of recession or depression in the first place.

Notwithstanding all of these potential problems it can be assumed that the British economy will continue to grow in the future and that it may start to move to a more stable growth pattern rather than the periods of boom and bust which have been so damaging in the post-war years. Whilst the nineteenth and early twentieth centuries saw periods of severe depression, this often affected only certain sectors of the economy and did not mean that the whole country was in a dire state. As mentioned in Chapter 2, when there was a depression in the agricultural sector, brought about by an abundant harvest which drove food prices down, the clothing and footware sectors of the economy boomed because of the extra spending power given to consumers. Such specific factors have, to a large extent, already been ironed out as the economy has become more interdependent. Households are not now just reliant on one sector of the economy for their employment and general well-being as all household members may well work in completely different fields. This is not true across the country as a whole as, for instance, the dependence of mining communities on the life of a coal mine makes abundantly clear. However, any isolated village or town represents a micro-economy which will become less and less common in the future, and therefore the desperation of the miners in their current plight will, hopefully, not be repeated so often.

Based on the use of a model for an economic cycle which represents a slow growth over time of some 2.2% in real terms, yet including periods of recession within periods of faster growth, a projected model of the economy, represented by GDP, can be constructed and is shown in Figure 13.2. The energy/GDP ratio produced from the data shown in the figure has been discounted over time by a rate equivalent to an energy saving of 1.5%/year. Although periods of recession are likely to lead to higher levels of energy saving and awareness this may not always be the case in view of the possibility that higher energy availability during periods of recession is also likely to lead

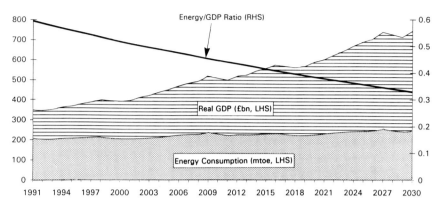

13.2 Forecast GDP and energy consumption/GDP ratio to estimate future energy consumption

to lower prices and, therefore, less incentive to reduce consumption. This appears to have been the case during the current downturn, although energy consumption is now, at last, starting to decline. The growth of GDP will also be related to the rise in population and therefore the energy/GDP ratio will reflect private energy consumption. Energy consumption in terms of million tonnes of oil equivalent (mtoe) is shown in Figure 13.2 and, together with consumption in terms of coal equivalent (mtce), in Table 13.1. Also included in the table are the projections of the share of the energy market accounted

Table 13.1 Predictions of energy consumption in the UK up to 2030

Year	1991	1995	2000	2005	2010	2020	2030
Total energy demand (mtoe)	207	205	203	216	225	226	244
Total energy demand (mtce)	364	361	357	380	396	397	429
Electricity share of demand (%)	15.9	18.0	20.5	23.0	25.5	30.5	35.5
Electricity share of demand (mtce)	58	65	73	87	101	121	152
Efficiency (%)	44.3	44.7	45.2	45.7	46.2	47.2	48.2
Total fuel consumption (mtce)	131	145	162	190	219	256	315

for by the electricity generation sector. This includes an assumption that electricity's share of consumption will increase by 0.5% a year. This is slightly higher than the 0.3% average annual increase over the past thirty years as higher petrol costs are expected to force people back to the railways or increase the use of electric vehicles. As the current generating efficiency is only in the region of 44%, after allowing for distribution losses, electricity consumption needs to be altered to take account of energy losses on conversion. This is accounted for in the final two lines of the table where efficiencies (assumed to increase by a modest 0.1% a year) and resulting fuel consumption are shown. All these figures are necessarily rough estimates, nevertheless it is hoped that they give a good approximation of energy consumption in the UK up to the year 2030.

WHAT WILL THE FUTURE BRING?

It is clear that Britain needs electricity both for its manufacturing industry and its service industries. The increasing consumption of energy, which is supposed to differentiate the advanced economies from the less advanced or advancing economies, means further that the amount of energy consumed around the world will increase and, unless a new and cheap method of producing energy is found, it will become increasingly scarce and increasingly expensive as the available supplies are used up. Many people state that there have always been new discoveries of oil and gas and that despite the fears that the world would run out of energy by 2000 it now appears that world oil and gas supplies will last until around the middle of the next century, at current rates of consumption.

However, it must be realized and understood that energy reserves, like everything else on this planet, are finite. Over the longer term there will be no more oil as the planet has not been in existence for long enough to create more than a certain amount through geological processes. Of course nuclear power offers a very long lasting alternative but it is unlikely that this will become the main source of the world's energy supplies in the immediate future because of the same environmental lobby which argues against the burning of coal. At least, in one respect, burning coal may have a lower environmental effect on the planet than nuclear power because of the continuation of the photosynthesis process which produced the forests from which the coal originated in the first place. Although the destruction of the forests as a result of acid rain and population growth may, itself, cause longer term problems.

A nuclear disaster would take centuries to clear up whereas a consistent and effective coal burning policy would have fewer deleterious effects. In this respect the introduction of a carbon tax of some variety should be welcomed rather than abhorred by the coal industry as it will enable people to burn coal with a clearer conscience than they might otherwise have had. This would be because the money raised from such a tax could be used both to retrofit desulphurization equipment to the existing coal-fired power stations and to undertake research and development into cleaner and better methods of burning coal which could increase efficiency to the mid-50% region. British Coal currently has its own research department but by all reports it is underfunded because the company is being forced to live within stringent guidelines laid down by the Treasury.

There are also problems of implementation and one which concerns FGD equipment is that the restrictions are founded on a country by country basis. This means that Scotland, which has more than enough electricity generation capacity for its own needs, will not need to fit FGD equipment on power stations situated within its borders. Scotland benefits from the relatively low sulphur content of indigenously produced coal although this will not mean that the country is free from any sulphur dioxide emissions, just that other parts of the UK require more urgent action. The country will therefore be at an advantage over England in the provision of electricity as most English power stations will require FGD equipment. It also bears little relation to the fact that acid rain knows only topographical and not human boundaries.

A further advantage of a carbon tax would be if some of the money raised was used in counteractive measures, and it is to be regretted that Norman Lamont did not use this carrot to offset some of the furore surrounding the removal of zero rating from domestic fuel in his March 1993 Budget. Such a programme could help fund a worldwide reforestation programme which would be an effective method of helping to ensure that the damaging effects, at least of the carbon dioxide production, were offset to some degree. It may even be that every company has to submit an energy or carbon dioxide and sulphur dioxide balance sheet as part of its annual report. In this instance a company would be in balance if it owned sufficient forests, etc to consume and regenerate the carbon dioxide produced in its own factories. If its carbon dioxide output was in surplus then the company could be taxed on that surplus with the proviso that the money would have to be used by the government in the creation of schemes to offset at least some of the damage which had been caused. Fluor and Enron in the USA both run an environmental programme along these lines and in the UK The Body Shop is a partner in a joint venture to build a wind generation plant in Wales with a capacity equivalent to the company's annual electricity consumption.

The developing world is very wary of any such incentives to offset the problems feared from global warming; however, such a programme could lead to investment in those countries because the industrialized countries would see that their land was relatively cheap and that the emission/consumption of carbon dioxide could be kept in balance relatively easily. As it would obviously have to be the case that every country in the world adopted such a strategy it would also help to offset some of the fears of the developing countries that the industrialized world is exporting its dangerous and chemically harmful industrial processes to these countries because of the lack of regulation and the relative ease of bribing local officials to turn a blind eye to polluting plants.

This has been particularly the case in the mining industry where it is not just the low level of wage costs and ease of extraction of ore which have attracted companies to areas which have had little mining in the past. In addition the low level of government intervention also helps to encourage them to invest because the smelters or refineries which they construct do not need to be built to the same environmental standards that would apply in North America, Europe or Japan.

The electricity conundrum will remain for the present but it is becoming clearer that coal will have a part to play in the world energy scene in the future and that there is likely to be a second coal age to beat the first coal age of the seventeenth and eighteenth centuries. What is particularly important, however, is timing. It would be cheaper to keep open existing mines if coal demand was to increase dramatically from, say 2020, than it would be to close these mines and then go to the vast expense of sinking new shafts and opening up new mines. If it was not until towards the end of the next century that coal started to regain its place as a source of energy then it would be better to close these mines and seal them up and thereby save the £3 million to £4 million annual cost of keeping them in a condition that would allow production in the future. Also, however, of supreme importance would be the relative cost of mining British coal compared with the cost of imports between now and the end of the next century. This is covered in the next chapter.

THE ENERGY BREAKDOWN

The past breakdown of energy supply in the UK is shown in Figure 13.3. The consumption of primary coal shown in the figure is only part of the total in view of the use of 80% of coal production in the generation of 60% of the electricity shown. Obviously coal has been supremely important in the past but is unlikely to be as important in the future in view of

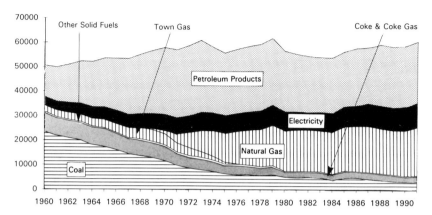

13.3 Final energy consumption 1960–91 (million therms)
Source: HMSO Digest of Energy Statistics, 1992.

this dependence upon the electricity industry for most of its demand and the moves within the electricity industry away from coal consumption and towards gas consumption.

The total amount of gas-fired power station capacity under planning or construction is, as present, equivalent to some 15.6 GW of output although only some 8.7 GW of this capacity will actually be brought to production over the next five years. As these plants are likely to be run in order to generate cash flow to pay back their capital costs it is likely that they will, initially at least, be run to the detriment of coal-fired capacity, even though they may be more expensive. And, although the regulator did not find the RECs in breach of their statutory requirements to purchase electricity as cheaply as possible the difficulty of breaking the long term gas supply contracts implies that they would have continued to use the plants in the attempt to generate cash flow. This would have been despite the financial penalties that the regulator could impose as most of the gas contracts are written on a take or pay basis and are site specific. This means that the consumers would still have to pay for the gas, even if they did not use it, and that they would not be able to sell it on to other consumers at other sites around the country. They would also suffer the interest costs on the use of the capital with no generation of income from the sale of electricity. Sir James McKinnon, the Chairman of Ofgas, said that this would be 'hard luck' for the generating companies but it was up to them to ensure that they could meet the conditions of the contracts themselves. If the government changed the rules that may be a different matter but he hoped that this would not be the case. As a result it was cheaper and easier to cancel or not renew coal supply contracts and gas has become the preferential fuel.

The other factor as far as the energy breakdown is concerned is that it cannot realistically be predicted as it is dependent upon the advance of scientific knowledge. The quest for a perpetual motion machine has consumed hours of experiment and still has to be discovered and although a scientific puzzle it will not provide a limitless supply of energy as such a machine would need to retain all of its energy if it was to keep moving. Any loss of energy and the machine would cease to function. Apart from this a large amount of work is being expended on natural sources of energy and on the possibility of generating electricity from fuel cells comprising a platinum/phosphoric acid battery which essentially combines hydrogen and oxygen to generate electricity and water. These cells have been in operation for a number of years, particularly in Japan, and they are gaining wider acceptance as their size and efficiency becomes more well known and the initial problems are ironed out.

The research into nuclear fusion as a potentially limitless source of energy continues apace. There has been a large amount of speculation as to whether or not nuclear fusion is a viable proposition and whether or not it will be controllable enough to prevent a major disaster from taking place. Such a disaster would be more likely to occur as a result of nuclear fusion at high temperatures in a large nuclear reactor operating as a controlled version of the sun. However, recent experiments seem to have shown the potential of using palladium as a catalyst to promote 'cold' nuclear fusion in the laboratory. As a result there is a growing belief that such a process would be possible to generate heat, and from this electricity, although there is still a large amount of scepticism as to whether or not the initially successful results were true or an elaborate hoax.

In the immediate future, up until the first quarter of the next century, it is unlikely that any new sources of energy will take up more than about 5% at most of the UK's generating capacity. This may well be a considerable overestimate as any advances cannot be guaranteed. Nuclear energy currently contributes some 20% of the country's electricity generation. This figure is likely to decline as the old Magnox reactors are decommissioned at a faster rate than it is currently anticipated that new generating capacity will be built. Indeed, it is interesting to note that the third part of the Sizewell complex, Sizewell C, has not been put in for planning permission despite the original intention that this would take place in October 1992.

The use of hydro-electricity is likely to grow in mountainous areas as it is both economical to run and the stored water can also be used to meet the demand for water from consumers. Whilst the lakes created take up large amounts of land and hence planning permission is likely to become more difficult, they are likely to continue to increase in the future, especially in

Scotland where Scottish Hydro was specifically set up to generate environmentally friendly electricity. In the future such schemes as the Severn Barrage are unlikely to be permitted because of the problems this would cause to shipping and navigation within the area. The generation of hydro-electricity outside Scotland and Wales is therefore unlikely to progress despite the increasingly environmentally aware nature of the population.

Finally there is the dirty trio: coal, gas and oil. The growth of gas generation has been covered in Chapter 12. Oil generation is likely to decline because of the probability that it will become increasingly expensive in the future. Despite continuing problems within OPEC the organization can be expected to improve its success in maintaining the price of oil and increasing it in the future. As a result oil can be expected to be used less and less in the deeply inefficient oil generation or cogeneration facilities with such plants moving to utilize coal in preference to oil. Oil is also a component in Orimulsion, which is a mixture of bitumen and water produced by Venezuela. It has an energy content some 10% higher than coal but also contains about 2.8% sulphur as opposed to UK produced coal's 1.6%. In the March 1993 Budget the Chancellor extended excise duty on oil fuels to include Orimulsion in order to comply with new European Community legislation. As a result it is expected that there will be a reduction in consumption from 1.1 mt of coal equivalent in 1991 to around 600 000 t in 1993 and succeeding years.

This reduction in Orimulsion consumption has been aided by the concern of Her Majesty's Inspectorate of Pollution (HMIP) which is considering the use of polluting fuels. This is particularly because the government was hoping that new gas fired stations would replace existing power stations and enable the country to keep to its commitment to reduce carbon dioxide emissions to the 1990 level by 2000. As it is it now seems that the 1990 level of 160 mt of contained carbon will be missed by miles as the Department of the Environment forecast in late 1992 that 2000 emissions will be of the order of 170 mt.

The methods which have been proposed to attempt to reduce carbon dioxide emissions include various incentives and are also likely to embrace some form of fossil fuel or carbon tax, on top of the current fossil fuel levy which is charged to help the nuclear industry cover its decommissioning costs. In this respect it is strange to see the amazing short termism of the government which relaxed the taxation on cars in the last budget, and again in the Autumn Statement, in order to help the UK motor industry during the recession. Nevertheless, the 1993 imposition of VAT on domestic fuel consumption is a step in the right direction but it seems that it will have little effect on reducing consumption. This is because domestic fuel use tends to be price inelastic and may only reduce carbon emissions by some 1.5 mt by 2000. As mentioned above conservation measures tend to be much more effective.

WHAT'S LEFT FOR COAL

The generation of electricity from coal-fired power stations is therefore expected to decline both because of the government's need to reduce carbon dioxide emissions and because the electricity generators and distributors will need to continue to generate cash flow to pay for the capital expenditure on their new gas-fired power stations. Whilst it may be cheaper for them to continue to operate their coal-fired generators and use this cash to pay the interest and capital on the borrowings on the gas stations this would probably require that they write off the capital cost of the gas generators in one large extraordinary item. As this could damage both their City reputations and their balance sheets the decision would probably be taken to stick with gas in preference to coal.

The move away from coal is likely also to be enhanced by the closure of old coal-fired power stations in the future and the decision not to replace them with new plants. Before the gas debate surfaced Derek Ezra argued that the maintenance on the use of coal

> would mean that active measures would be required to maintain the effective coalburning capacity at somewhere near its present level of around 40 GW [in 1982]. As very little new coal-fired capacity has been installed during the last decade, this will require the refurbishment and reconstruction of existing coal-fired stations as they become obsolescent, or by the building of new coal-fired plant using the latest coal combustion technology.

Despite the closure of some stations in recent months it is likely that they will be dismantled rather than be kept in working order to be brought back into production if required by an increase in future UK energy demand. As such coal's share of the electricity generation market will show a relative decline.

The use of French electricity provided through the undersea cable is unlikely to have much of an impact on the coal generating capacity. This is because it can be expected to continue to supply electricity in the future to the level of the 6.5 mtce which it currently supplies, even if it is forced to reduce its price by 11% to comply with any Monopolies Commission investigation. The reason for this is that France retains a massive generating capacity surplus which was brought about by its massive nuclear power station programmes instituted in the 1960s and 1970s. France still needs to dispose of the excess electricity and despite the need of Italy, Portugal and Spain to import electricity in view of their lack of capacity it is likely that France will still need to sell electricity to the UK. Indeed, France's excess generating capacity would surely mean that it would be able to increase its own supplies to Italy,

Table 13.2 Projected fuel consumption in electricity generation (mtce)

	1991	1995	2000
Demand	131	145	162
Coal	85	70	72
Oil	10	5	2
Gas	0	30	52
Nuclear	27	32	30
Hydro	2	2	2
Total	124	139	156
Imports	7	6	6
Total supply	131	145	162

Portugal and Spain rather than allowing the UK to turn around the flow of electricity through the Channel link (it currently exports around 12% of its local output). In any case National Power has joined a consortium to build a new coal fired power station in Portugal. This station will use imported coal which is very unlikely to be of British origin.

Table 13.2 is based on the figures in Table 13.1 for total fuel consumption in electricity generation. It shows that increased gas generation capabilities will more than offset the rise in fuel demand for electricity generation, assuming that all 15.7 GW of plant currently planned or under discussion is brought into operation. And, if such a buoyant scenario for electricity demand fails to materialize it will mean a reduction in coal rather than gas consumption in 1995. This is because gas and nuclear electricity production will account for almost all of the 25 GW of base load demand and that coal will be squeezed out of this area of the market. Whether or not coal is supplied from indigenous sources or from imports is the subject of the next chapter.

Chapter
14

Home production v imports

A ONE-WAY STREET?

The dramatic change in political philosophy in 1979 did not, initially, cause British Coal much of an upset. Indeed, the Corporation had been travelling down a one-way street of redundancies and pit closures since nationalization (see Part I). Even the abolition of coal import restrictions, one of the acts of the Conservative government's first parliamentary session, had little immediate effect (Figure 14.1). It was only in 1984/5, when the year-long strike forced consumers to look elsewhere for supplies, that imports of coal started to become a serious threat to the indigenous industry.

However, the ability of British Coal to meet major inland demand meant that overseas suppliers were severely restricted in their attempts to break into the local market. Previous import restrictions, high prices and problematic local supply, due to the militancy of the mining unions, had also encouraged many companies in the private sector to move away from coal. The scope of the non-generation market was therefore much reduced even before the 1979 General Election.

As a result British Coal became ever more dependent on the cosy relationship that existed between it and the CEGB. And, as both companies

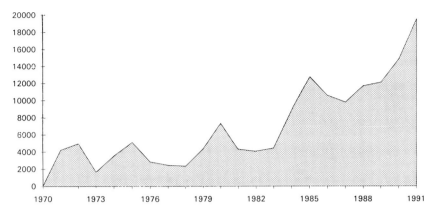

14.1 British coal imports 1970–91 (000t)

were in the State sector its expenditure on sales and marketing was significantly lower than other similar raw materials companies. This, in turn, must
have reduced British Coal's ability to win new markets within the UK and left
it largely reliant on the privatized descendants of the CEGB: National Power
and PowerGen.

And British Coal cannot claim to be without the benefit of hindsight for it
has already lost two markets in recently privatized companies. First, there was
British Steel which, together with the other UK steel producers, now imports
over 90% of UK coking coal consumption. This compares with the much lower
8% of other coal demand which is met from imports. The second area where
British Coal has lost out to imports is in the Scottish electricity generation
industry. Here almost 50% of the coal fired generation market is supplied
by companies other than British Coal, and a fair amount of this is from
imported coal (see page 192). Unlike National Power and PowerGen, Scottish
Power was able to negotiate more favourable purchase contracts in advance
of privatization, although they are not due to be renegotiated until 1995.
Nevertheless, the lower amount of indigenous coal taken by the company must
indicate the future level of demand across the UK as a whole.

The need to ensure a free market in the supply and demand for coal was
one of the central themes of the DTI Coal Review. And the definition of the
free market was taken to be a free international market, otherwise British
industry would be handicapped by high local power costs against much lower
costs elsewhere in Europe. In actual fact the UK should have an advantage as
far as the cost of energy supplies is concerned because of the country's
situation on some of the world's main trade routes. On top of this the abundant

local supplies of coal, gas, hydro-electricity and oil should combine to reduce the local cost of power in a truly competitive market. Because of the many problems over ascertaining its true cost it should, of course, be noted that nuclear power is excluded from this list.

However, unfortunately, the 'Buy British' lobby conspired to restrict inports of coal and, as a result of a seemingly guaranteed market, the price competitiveness of British Coal became unimportant and declined. The other problem facing a 'Buy British' lobby is that it fails to recognize that coal is a wasting asset and that the more that is produced today the less there will be available for the future. Admittedly, the total amount of coal still underlying the UK could, if strict economics were ignored, meet demand for over 300 years. And, although production has taken place on an organized basis for at least 800 years it is only over the past two centuries that it can be considered significant. There is, therefore, in historical terms, not necessarily that much coal left – especially if it becomes an important fuel again when the oil runs out – and it surely makes political and economic sense to buy cheap production from overseas today in order to ensure the ability to produce locally in the future if scarcity and high prices warrant.

Under the current administration this is a decision left to the market, and the extreme scenario outlined above is beyond its scope. It is therefore up to the government to ensure that a precious resource is not squandered in the attempt to make a substantial profit one day at the expense of security the next. Sir Denis Rooke, obviously really a champion of gas but still a defender of the country's total energy supplies, could not have put it better when arguing in favour of a concerted energy policy which would help to guarantee the country long term supplies of fuel. He stated that:

> any additional sales of gas into non-premium markets are likely to be made partly at the expense of coal. The consequent fall in short-run coal demand could impair the coal industry's ability to meet longer term requirements. Thus the more rapid depletion of natural gas reserves could on the one hand bring forward the need for additional coal for making SNG, while on the other hand increasing the difficulties of meeting the longer term demand.

Admittedly, he was writing at a time when the level of known reserves of gas in the North Sea was lower than today, and the extent of geological knowledge was such that exploration successes could not be guaranteed. Nevertheless, it would, today, be very easy to be lulled into complacency in the expectation that because more reserves were found yesterday they will be found tomorrow. Any extractive industry is governed by the law of diminishing returns. The more that is produced the less there is to be produced or discovered in the future. Assuming that the price of the product remains constant the profit made on extraction will also tend to decline over time (the invention of new technology notwithstanding) as a company will attempt to generate the highest level of profit possible in any one year. It will obviously have regard for its longer term future but if there is a choice between two gas fields of similar size and quality the one which is cheapest to exploit would normally be chosen first.

SAFEGUARDING MINERAL ASSETS

Great Britain has had little experience of safeguarding national mineral assets. Whilst mining legislation has evolved over many years, State ownership and hence control of mineral extraction was for a very long time restricted to gold and silver. State ownership of these metals was required because of their importance in helping to fund government expenditure. This limited ownership of minerals was extended in 1942 when the coal royalties were nationalized and all of Britain's coal was vested first in the Coal Commission and, subsequently, in the National Coal Board. With no precious metal extraction the State, except through the quasi-autonomous and deeply restrictive activities of British Coal (aided by out-of-date legislation), has no effective control over the use of national mineral assets and little direct interest in whether or not they are being mined to the long term benefit of the nation.

At the time of the nationalization of the coal royalties the government already effectively controlled the industry and so it made little difference to the status quo. However, there remains little control on British Coal to ensure that it makes the best use of the assets it is entrusted to manage. All of the coal is vested in the Corporation rather than in the state and it is therefore up to British Coal to decide whether or not to grant an operating licence either to itself or to any of the independent operators.

Further, there is no control on the day-to-day management of British Coal beyond its ultimate accountability to the DTI, and hence to Parliament. This means that British Coal is able, quite freely, to close down pits without having to take into account the long term needs of the country's energy supplies, or whether another operator could produce coal more cheaply and thereby increase the level of economically recoverable reserves. This would have the side-effect of extending the life of more valuable and easily extractable reserves and reduce the propensity for high grading which many people fear. In his evidence to the Parliamentary Select Committee on Trade and Industry Neil Clarke admitted that the pits on the list of those to be closed (he refused to call it a 'hit list') changed over time. He was therefore forced to close a pit in one year that could make a profit in the next, although the ability to make a profit may have been conditional on the pit receiving artificially high prices for its output.

Part of the reason for this was British Coal's inadequate planning (perhaps at the behest of government departments which did not want to see pit closures ahead of the 1992 General Election) which resulted both in significant over-capacity and a massive build-up in stocks and, therefore, forced the company to find convenient reasons to close the 31 pits on the original list. Nevertheless, the control of mining operations to prevent closures on short term economic grounds, especially as a result of low cost 'rape and pillage' operations at competing collieries is a potentially serious problem which the government must address.

The best way to address the problem is to adopt a system similar to that used in South Africa to ensure that only the lowest grade of gold ore to make a profit is worked. In this way the South African legislation makes sure that there is no high grading of gold deposits which would leave gold in the ground that would be profitable were it to be mined in conjunction with other, higher grade material. As such the country is able to ensure that its reserves and resources are mined in the most economic fashion and that the state will obtain the highest benefit both to its balance of trade (as the majority of the gold produced in the country is exported) and to its tax income because the more gold that is produced from the country's mines the more tax revenue will be generated.

Such a system would be difficult to implement exactly in this country because coal is not mined in an ore but rather it makes up the whole of the rock that is extracted. Nevertheless, some use of a cost of extraction calculation in the creation of a lease or royalty would make clear economic sense. Indeed, it should also be applied to the oil and gas industries because it would then ensure that all of the gas and oil which can be extracted at a certain price will be extracted at that price. The problems produced by the dash for gas would not recur because they are only made possible by the availability of cheap gas to the electricity generators. This, in turn, has only been possible because the gas producers are likely to make large profits on the supply of this gas and may also be exploiting highly profitable reserves at the expense of the country's long term energy interest.

Such a royalty would also make the coal, gas and oil producers think twice before they extracted the energy source from inside this country rather than supplying it from imports. As a result it may well lead to a longer life for the country's wasting assets than is currently envisaged, and would therefore avoid the horrific scenario painted for around 2030 when Britain could run out of its own indigenous fuel sources and would be dependent upon the outside world for its power.

Much has been made of the relatively free market in world energy and the supposed impossibility of an energy war breaking out in the future. What it supposes is that new sources of energy will become available in the future, whether by invention or by the discovery of an endless supply of fossil fuel. The first may be a pipe-dream; we may never know. The second is impossible. As such the world's coal producers will once again rule the world and it is at this time that to have indigenous supplies of the fuel will provide Britain with a great advantage. Supplies is an important word because there is no point in owning vast coal resources if they cannot be extracted economically because of the legacy of the mining industry of the past. In such an eventuality Britain would not be dependent on stable countries such as Australia or the USA for its coal supplies as these countries would be expected to use most of their output internally – indeed there may even be a ban on exports. Britain would then have to turn to countries such as Colombia or Venezuela and would have to compete with the rest of Europe for all that these countries would want to sell and at the price they want to sell it. It is therefore clearly necessary for Britain to retain an indigenous expertise and industry in coal extraction rather than being left to the whims of what must be considered to be relatively unstable political regimes.

FUTURE SOURCES OF SUPPLY

Although the increase in gas generation capacity over the next few years is a major factor in the decline in demand for coal in electricity generation (see Chapter 13) the high price of British produced coal remains important. It is this price which has encouraged consumers to look elsewhere for an increasingly large level of supplies in order to reduce costs. As mentioned in Chapter 9, the UK imported 10.3 mt of steaming coal in 1991, and some 7.3 mt of this was used by the electricity generators. This figure was expected to rise substantially following the March 1993 expiry of the coal supply contracts despite the desire of the two big generators to run down their massive coal stocks. However, the government's decision to subsidize British coal production (over and above the 40 mt in 1993/4 and 30 mt in 1994/5 agreed in the new supply contracts) to ensure that it could compete with imports altered this assumption. As a result an additional 12 mt of indigenous demand was anticipated which would maintain 12 further pits in production. Whether or not British Coal is able to obtain any extra sales still has to be seen and, unless real costs of production are reduced such intervention can only be a short term placebo.

There are two reasons for the continuation of the coal industry's problems. First, both National Power and PowerGen have invested in large coal import facilities and will want to see some return on this investment. Further, they would be able to switch to imported coal with little difficulty should the cost of production not be reduced quickly enough to make British coal world competitive in two years' time. These new import facilities are expected to become fully operational during 1994 at which time total coal import capacity will be of the order of 30 mt. It will therefore be more than able to meet generating demand, whilst it will also facilitate an increase in coal imports for use in other areas.

Secondly, British Coal's high sulphur content in comparison with internationally traded coal should lead to a natural reduction in demand as environmental regulations increase. And whilst Scotland does produce low sulphur coal it must now be regretted that the UK's other low sulphur area, the Kent coal-field, was eventually forced to close as a result of high costs and union militancy. Further, there seems little point in the generators burning more high sulphur coal than they need to and the obvious solution would be to consume their stockpiles over the next two years. This would leave them free to purchase low sulphur imported coal in the future and would have the added advantage of releasing cash for other working capital purposes. Indeed, both generators threatened the government with a stockpiling charge (in the case of PowerGen this was in excess of £140 million over five years) if the White Paper restricted them from utilizing their stocks.

Taking the figures for UK coal demand in electricity generation from the last chapter the breakdown across regional areas is expected to be as shown in Table 14.1. It should, however, be remembered that the forecast for 1995 estimates a total fuel burn some 14 mtce greater than 1991. If this does not materialize then coal, as the swing supplier, will suffer.

It can safely be assumed that transport costs will prevent British Coal from selling to Northern Ireland, whilst the removal of the Corporation's 'all or nothing' contracts may reduce consumption by other generators by about 50%. In Scotland the future shape of coal supply contracts is already in the public domain. Whilst 2.5 mt of the 4.8 mt consumed by Scottish Power annually is supplied by British Coal on a long term contract which expires in 1995, the supply of a further 0.5 mt has been agreed on short term contracts. Of the remaining 1.8 mt specific contracts covering about half this have been signed with local private producers (100 000 t), Australian producers (360 000 t), Polish mines (250 000 t), the CIS (170 000 t) and the USA (60 000 t). Finally, British Coal's base contracts with the generators will account for 30 mt of 1995

Table 14.1 Regional breakdown of UK coal demand for
electricity generation

mtce	1991	1995	2000
Total consumption	85	70	70
England and Wales	75	60	60
Scotland	5	5	5
Northern Ireland	1	1	1
Private generators	4	4	4

Table 14.2

mtce	1995
Generator coal consumption	70
England and Wales base contracts	30
England and Wales open cast	10
Scotland Longannet	2.5
Scotland private and imports	2.5
Northern Ireland imports	1
Private generators British Coal	2
Private generators other	2
Stock run-down	5
	55
Uncommitted	15
	70

demand. These figures are shown in Table 14.2, together with an assumed
figure for stock run-downs to leave a figure for potential, uncommitted demand
in 1995.

In addition to this uncommitted figure, coal consumption in other areas of
industry is likely to be similar to current demand which, after accounting for
exports, is about 14 mt. There is therefore a potential additional market of
29 mt of which 5 mt will be supplied by other open cast and private production
and around 10 mt from existing levels of imports. This leaves some 14 mt of
net uncommitted demand which was originally expected to be met from
imports. Following the White Paper it now seems that this figure will be met

from subsidized indigenous supplies, although it remains up to British Coal and new private operators to ensure that costs are reduced to a level that enables them to retain this market in 1996 and beyond. However, it must be stressed again that this 14 mt only exists if electricity consumption increases along the lines forecast in Chapter 13, as without such an increase there will be no scope for a rise in coal demand.

THE LONGER TERM POTENTIAL

Any sensible strategist would want to introduce as much diversity into a system as possible in order to prevent the system from being compromised by events outside its control. This is the reason why Britain moved towards oil as a fuel in the post-World War II years. This strategy was clearly ill advised because it meant that the country became dependent upon a few suppliers who were able to control the market by virtue not of an effective monopoly but of the ability to act as a swing producer and push the world into an energy deficit. This spurred the industrialized nations to look for oil supplies elsewhere and their success had the effect of reducing the Arab dominance of the oil market. It must, however, be remembered that this dominance was maintained for many years, and the Arabs were still able to force up the price of oil by a considerable amount in the 1979 oil shock, a full seven years after the Yom Kippur War.

Everything takes time to evolve; it takes time to decide where to explore, to carry out that exploration and then to carry out the necessary economic feasibility studies to determine whether or not a specific investment is profitable. Whilst an individual company or corporation does not need to take this into account, as it has a first duty to provide its shareholders with as much profit as possible, the good of the nation, unfortunately, is put into second place. It must be the role of the state to legislate to ensure that the companies and individuals within its boundaries are not pursuing short term ends which will adversely affect the long term needs of the country.

Such legislation goes against Conservative free market/laissez faire economics, and it must be remembered that the State cannot be guaranteed to be accurate in its forecasting, as can be seen from the problems resulting from the over-dependence on oil in the 1970s. But despite this incorrect assumption in those years of cheap oil it was still maintained that the British coal industry had an active role to play in the future energy supplies of the country. This was not just because the government during the latter part of the 1960s was Labour, under Harold Wilson who first made his mark as a wartime

194

civil servant in the Department of Energy. And not just because the Labour government was committed to retaining as much of the coal industry as possible. Indeed, more pits were closed under Labour administrations in the latter 1960s than by the Conservative administration of the early 1960s. The belief was quite simply that Britain's future depended upon its ability to obtain relatively cheap and secure energy supplies and a major source of that supply was coal.

The discovery of oil in the North Sea did not change that policy markedly as the price of oil was still totally dependent upon the whims of the major producers in the Middle East who could, effectively, control the market. This is a situation which must not be allowed to recur and can only be achieved if there is a structured energy policy which, such as the royalty or lease based system outlined above, promotes the longevity of the country's fuel supplies. It is not necessary to ensure that coal is produced today or tomorrow; it is necessary to ensure that it is possible to produce coal cheaply and efficiently today or tomorrow, having regard to the situation of the general energy market around the world at the time. Few forecasters take economic cycles into account in their projections and the current Trade and Industry Select Committee has castigated previous forecasters for being hideously wrong in their projections of world energy demand and world energy prices in the past, when the actual prices we see today are considerably lower.

These forecasters are unlikely to have taken into account the possibility of a recession in their forecasts, believing implicitly that governments would be able to achieve all of the economic goals put forward in their election manifestos. As a result world energy demand is significantly lower today than was previously forecast. However, this does not mean that it will be significantly lower in 1997 than is currently forecast. Indeed, it is likely that it will be significantly higher. As the economic cycle turns upwards more energy will be consumed and the initial boom from the bust which is currently being experienced may be considerable. This, of course, is still dependent upon the world leaders being able to engineer a recovery which, as some commentators still fear, may falter and lead to a more massive collapse along the lines of the depression of the 1930s.

In such a dire situation the UK will become even more dependent on its indigenous fuel reserves because of the reduced ability to import fuel. This would be seen through the imposition of trade controls and restrictions, like those indicated by the Clinton administration soon after the President's inauguration in early 1993, and by the introduction of foreign exchange controls in order to attempt to maintain the currency at a level which does not lead to the hyper-inflationary levels which surfaced in Germany during the inter-war years. It is therefore a mistake to close down all of Britain's collieries, but it is also a mistake to keep all of them open in the belief that it will maintain the country's energy independence in the future.

Finally, of course, and more in keeping with the government's free enterprise philosophy, it should open up every mine in the country (not just those on the closure list) to a competitive tendering whereby each interested party could bid what it thought the mine was worth, the mine then going to the highest bidder. British Coal should receive the benefit of any bids which were in excess of its own and this could then be used to repay the government for any outstanding loans which still remain. This would let the free market decide the country's future energy requirements with regard to either home production or imports and the methods and techniques of such a sale are covered in the next chapter.

Chapter
15

Methods of disposal

BACK TO BASICS

The nationalization of the British coal mining industry was meant to improve its international competitiveness. Up until this time many of the mines had been in individual hands and there had been no opportunity for the rationalization and combination which would increase productivity and output. As only the larger mines were nationalized this left a number of small mines employing less than 30 men underground and this legacy remains today. There are now only 142 mines employing a total of 1647 men underground in the private sector compared with 480 mines at nationalization.

As a result of the legislation restricting the size of the underground mines under private control Britain does not have a private sector with much experience of large scale underground coal mining. And many of the companies which hope to move in and take advantage of British Coal's sell-off have not been able to survive the time needed to wait for the industry to be privatized. Indeed, it is a peculiar absurdity of the free market Conservative administration that it has continued to restrict the number of men working underground and left the ownership of the country's coal within the Corporation. This means that it remains up to British Coal to determine not just how, when and who should mine any particular seam, but also the level of royalties which a private operator would have to pay for the privilege. As coal royalties remain a considerable cost British Coal has been able effectively to restrict new entrants and competition.

Indeed, the worries about the possibility that British Coal could abuse its position even prompted a referral to the Monopolies and Mergers Commission (MMC) in 1983. Although not wholly happy with the situation the MMC stated that at least the few private operators that did exist would be able to provide some competition and comparison with the NCB's operations. But because the output from these mines is small in comparison with British Coal's production the Corporation has still been operating in an effective monopoly, especially with its contracts to supply the electricity industry in place until April 1993. It may be that the Corporation did not envisage that it would actually face the prospect that its contracts would not be renegotiated with the electricity generators and that they would attempt to flex their newly found muscle to the detriment of the coal mining industry. Nevertheless, this is the situation which British Coal currently has to face, and the sudden rationalization forced on it in October 1992 smacks of panic rather than of carefully thought out proposals to improve productivity, profitability and viability.

Strangely, the company did not seem to object when the initial contracts were drawn up with the electricity generators in advance of the generators' privatization. Indeed, the company is almost certain to have stated that it was happy that the contracts offered security to the miners and their families over the coming years. This, however, could only have been the case if the Corporation was able to continue to reduce costs to a level at which it would have been able to match international prices. In the event this has not been the case.

A large part of the problem, however, is not of British Coal's making. And, despite some political protestations to the contrary, the blame cannot be laid at the feet of Arthur Scargill or the NUM. Specifically the blame rests with the great expansion in cheap open cast coal mining capacity around the world and the recession which has caused a reduction in fuel prices because of a fall in demand for energy. It is because of this that international coal prices have fallen and it is because international coal prices have fallen that British Coal is not currently competitive in the international market. This factor needs to be taken into account when the government proceeds with the privatization of the industry.

Unlike the other nationalized industries which were undeniably heavily overmanned, British Coal will not be privatized with a large workforce and then allowed to reduce its workforce over time and thereby enhance profitability. The poor state of the company's finances, and the need for massive government assistance with respect to the redundancy programme, are obviously the main reasons behind this decision. As a result the Corporation could now be sold with little need for a massive redundancy programme and all of the associated problems which that would cause were the industry in the private sector. This is not to say that there will not be any further rationalization which can be carried out when the company is returned to the private sector, by whatever means is finally chosen, but it means that most of the politically sensitive areas will have been covered by the government rather than being left to be suffered by British Coal's successors.

A SPECIAL CASE

That British Coal is a special case can clearly be seen from Part I of this book. In many respects it is more closely allied with British Leyland than any of the 46 other industries which have been privatized. This is because of the number of employees – even allowing for the closure of 12 pits and mothballing of six more, there will still be some 30 000 people left in the industry – and because of the strategic importance which is placed on the industry. Apart from this the large number of union members and the militancy of the union means that there cannot be much done without obtaining at least a modicum of union blessing. That the redundancy programme was possible obviously shows how demoralized the union is, but in many ways they were faced with a *fait accompli* and they were threatened that any strike action would lead to the removal of redundancy payments. As these payments added up to a considerable sum there were very few union members who wanted to press for strike action in order to defend their jobs. Apart from this, the failure of the 1984/5 strike to achieve the aim of preventing pit closures means that it would have been a battle which few union members would think they had any chance of winning.

The other problem as far as the sale of the industry is concerned relates to the complicated structure which has grown up since nationalization in 1947. The Corporation is now split into five regional boards which, together with the Opencast Executive, control the mining operations. The selling is controlled from the head office at Hobart House in London where most of the planning and forecasting functions are situated. As a result the Corporation does not have much in the way of intermediate management around the country which could take over the operation of the mines if they were sold off by taking an arbitrary decision to split the company into a number of operating segments. The mines would therefore need to be sold into organizations which have an existing management structure which would immediately be able to co-ordinate the running of the business.

One of the main problems here would be that if the company was split up after a number of coal supply contracts had been negotiated with the electricity generators they could either be split pro rata by the mining capacity of the operations in their post-split form (as occurred with the Electricité de France supply contract at the time of the vesting of the RECs), or they could be split along the lines of locality, on a pit by pit to generator by generator basis, leaving the residual pits to fight amongst themselves for market share. Or the division of the pits between companies could be carried out in such a way that each company had at least one pit in any specific area and the contracts could then be split on a locality basis with each pit in a specific area being allowed to produce some contract and some spot coal.

Coal could not have been privatized in advance of these contracts being completed as it would still be suffering from the overmanning which the government decided it was its responsibility to address in advance of privatization. This is because of the large costs involved and the outcry that would have come from the government's supporters if they had bought into the industry and then found that there was little of the industry left for them to attempt to make money from. The trouble that the government faces as a result is that the sale of the electricity industry has effectively tied its hands as far as the coal industry is concerned and there would be little sense in splitting it up into too many small parts even though this would lead to the greatest amount of competition and improve the lot of the consumer of coal, and hopefully of electricity.

The further problem arises with what to do with the Opencast Executive. There is absolutely no sense in splitting it up into various units. British Coal is due a royalty from the operators of the pits and it is able to put this income to good effect. If it was suddenly taken away from the Corporation then it would not be able to operate effectively as much of this income is needed to pay interest to the government on its borrowings and cover administrative expenses. As it appears likely that the 6 mothballed pits will not be included in any initial phase of privatization it would make sense if the Corporation could retain the royalty income in order to offset the care and maintenance costs. Whilst this is likely to cost up to £24 million a year it is possible that there will be a surplus which could also help with the development expenditure at the Maltby colliery.

As far as future royalties are concerned it would not be sensible for British Coal to remain in control of all of the coal in the ground and, as already mentioned, it would be sensible if this was vested in a separate authority which could license areas for either open cast or underground mining much in the same way that the energy section of the Department of Trade and Industry now auctions off blocks for exploration and exploitation of oil and gas in the North Sea. In this instance the principle of the free market would be maintained and enhanced and the government would reap some direct revenue from the royalties when and if new mines are constructed. This income would pay for the organization of the department and for the necessary health and safety department which would probably need to be expanded when the majority of the mines have been moved into the private sector. Indeed, as a large part of this work is currently carried out by British Coal there would be some savings for the company which would increase its attractions as far as the private sector is concerned.

THE SPECIFICS OF SALE

Tables 15.1–15.4 list the pits which currently exist in the industry. Table 15.1 includes all of the pits which will continue in operation whether or not the government continues to subsidize the industry beyond 1995. Table 15.2 includes the 12 which were given a reprieve whilst Table 15.3 shows the six pits to be placed on care and maintenance and Maltby where development work is to be undertaken. Table 15.4 shows the 12 pits which are now set to close. The fourteen pits on the Rothschild list are marked with an asterisk.

As far as the timing of the sale is concerned and how it should be carried out, it would make considerable sense if it was as transparent as possible in order to satisfy all of the critics of government policy that exist in the coal industry. As previously mentioned the sale should be preceded with government legislation which would lead to the setting up of a new department within the DTI to administer the newly privatized industry, although as was the case with British Steel there should be no need for a regulator to ensure that the price of coal was not being artificially maintained in order to boost mining company profits. Indeed, the availability of relatively cheap imports of foreign coal have shown that there is no need for this expense, whilst the MMC would be able to investigate any suggestions of restrictive practices. The legislation should also include the basic removal of the constraint on private ownership of coal mines employing more than 150 miners underground and on the annual output of private mines.

This legislation must be passed, or at least guaranteed, before any sales can take place. This is necessary because at present the government has only given an undertaking to look at the legal position of the independent mines. It has not, so far, confirmed that they will be allowed to operate on the same basis as British Coal and, therefore, independent producers may be dis-

Table 15.1 The pits likely to be left in operation whether or not subsidies continue beyond 1995

Area	Pit	1991/2 output (000t)	First half 1992/3 (000t)	Sulphur (%)
Scotland	Longannet*	2104	928	0.44
North-east	Ellington	2066	1057	1.22
North Yorkshire	Stillingfleet*	2017	905	1.15
(Selby Group)	Wistow*	2078	1306	1.15
	Whitemoor	1055	692	1.15
	Riccall*	2149	956	1.15
	North Selby	1102	851	1.15
North Yorkshire	Kellingley*	2007	879	1.78
(Other)	Goldthorpe	1442	721	1.80
South Yorkshire	Silverwood	1444	489	1.47
	Manton	1124	728	2.15
Central region	Littleton	1200	617	1.26
	Daw Mill*	2311	1264	1.35
Nottinghamshire	Annesley/Bentinck	1360	696	1.68
	Harworth*	1977	915	2.31
	Thoresby*	2222	965	1.81
	Ollerton*	1600	889	1.83
	Welbeck*	1574	724	1.96
	Asfordby	Not yet in production		
South Wales	Tower/Mardy	694	383	1.00

couraged from purchasing any large mines which are offered for sale. This is simply because a large mine needs to be operated at maximum output in order to reduce the unit cost of all associated overheads.

In the initial phase the government has indicated that the 12 pits on the closure list will be sold. This may seem to be sensible because it will reduce any refurbishment costs which will have to be spent before production can resume. Unfortunately, this timing of such a sale will cause two problems. First, these mines will find it difficult to break into the steam coal market (even with the government subsidizing output down to free market levels) because of the advantage given to the 12 additional pits that remain in production. Second, there may be few buyers for these accepted high cost collieries because the purchasers will want to wait for the more attractive pits

Table 15.2 The 12 reprieved pits

Area	Pit	1991/2 output (000t)	First half 1992/3 (000t)	Sulphur (%)
North-east	Wearmouth*	636	254	0.67
North-west	Point of Ayr	556	213	1.91
	Silverdale	935	510	3.09
North Yorkshire	Prince of Wales	1594	952	1.38
	Frickley*	1054	494	1.94
South Yorkshire	Bentley	1099	558	1.67
	Hatfield	595	368	1.13
	Kiveton	1103	432	1.95
Central region	Markham	1923	758	1.72
Nottinghamshire	Bilsthorpe	873	692	1.67
	Rufford	981	566	2.03
	Calverton	860	539	1.14

Table 15.3 The six pits to be placed on care and maintenance and Maltby

Area	Pit	1991/2 output (000t)	First half 1992/3 (000t)	Sulphur (%)
South Yorkshire	Maltby*	1315	638	1.49
North-east	Westoe	1468	492	1.46
	Easington	1443	657	1.58
South Yorkshire	Rossington*	1061	331	1.24
Central region	Shirebrook	1399	683	1.80
Nottinghamshire	Bevercotes	1270	447	1.09
	Clipstone	973	406	1.68

which are not due for immediate closure and are expected to be sold off in 1994/5.

In view of these problems all of British Coal's pits over and above the 20 which were due to remain under the Corporation's control should be put out to tender at the same time; including those on the mothballed list and Maltby. Obviously a minimum tender should be fixed for each mine, which should be

Table 15.4 The twelve pits set to close

Area	Pit	1991/2 output (000t)	First half 1992/3 (000t)	Sulphur (%)
North-east	Vane Tempest	832	259	0.81
North-west	Parkside	871	376	1.37
	Trentham	2335	491	1.18
North Yorkshire	Grimethorpe	947	411	2.66
	Houghton Main	380	220	2.66
	Sharlston	1014	438	1.79
South Yorkshire	Markham Main	832	265	1.20
Central region	Bolsover	650	295	2.26
Nottinghamshire	Silverhill	932	359	1.67
	Cotgrave	902	272	1.70
South Wales	Betws Drift	126	77	0.64
	Taff Merthyr	634	268	0.60

broadly equivalent to the sum of the recent capital expenditure on the mine, say its next three years' projected earnings and a nominal sum dependent on the level of estimated recoverable reserves. If this figure is not exceeded then the mine should remain in the control of British Coal and be closed if it was on the closure list. The bulk of the information required in calculating the minimum tender should be made available to potential purchasers, although not the actual minimum price. In this way the company should be able to raise as much money as possible. In view of the huge amount of argument which surrounded the application for television broadcasting licences there should be no caveats which would favour individuals or companies which could show a better industrial or other record as such an allowance would reduce the sale to an arbitrary decision which would not be in the best interests of the mines, the miners or the markets. Nevertheless, it would be necessary for the bidding company to show that it would be able to finance the acquisition and to show that it had enough working capital to continue to operate the mine over at least the succeeding two years.

All of the money obtained by these sales should be used by British Coal either to repay the government for any outstanding borrowings (£691 million at the end of March 1992) or to carry out the capital works needed to improve the long term productivity and profitability of the mines left under its control.

This, indeed, should give the company the incentive to obtain as high a price as possible for the mines. It may also influence the Corporation to keep as many mines open as possible as this would lead to an increase in the total amount of money which the company could possibly be able to raise and therefore reduce the level of the company's borrowings by the maximum possible amount.

Following the completion of the sales British Coal would have to be looked at again. If it was considered that the income which could be generated from the remaining mines and from the open cast royalties would be sufficient to enable the company to remain in profit and to increase its profitability in the future – especially as a result of the reduction in borrowings following the repayment of the government loans – then it may be possible to proceed with a flotation on the stock market. Otherwise the government may wish to write off any further outstanding loans and liabilities as these would, largely, have been incurred as a result of the need to continue with capital expenditure and redundancy programmes at the government's behest rather than in the proper and profitable management of an independent enterprise.

If such a sale is not considered, then the rump of the industry could be offered in a final sale to the interested parties which had made offers for the mines in the original tender offer. In this way the government should be able to maximize the amount of money it raises from the sale of the coal industry, although it should attempt to sell the final rump of the mining industry on the stock market rather than in a trade sale to one of the interested parties. This is because of the competitive advantage which an individual company would obtain in any such sale and because of the government's stated intention to increase the amount of public ownership of the previously nationalized industries.

The five mines within the Selby complex would probably be sold in a single unit, either to a single buyer or to a joint venture between two or more buyers. Indeed, with a turnover of some £264m/year they represent a significant company in their own right. As the operation would control most of the coal production from the area there may be few takers for the other mines in the North Yorkshire region and some of these may remain within the confines of British Coal in its post-nationalization form. British Coal may also be left with the Ellington/Lynemouth Colliery in the north-east region whilst the South Yorkshire, Central and Nottinghamshire regions are likely to find buyers. Ryan Group, which is active in the South Wales area, has shown interest in both the Betws Drift and Taff Merthyr Collieries.

The rump of British Coal would also have the advantage that it would be possible for the company to increase its potential profitability through a reduction in its central overheads which would enhance its interest to the general investing public. As already mentioned some of the employees would transfer to the DTI to work on the licensing of new open cast and underground coal mines across the country. In addition, the central overheads could be reduced by a reduction in the numbers of employees who are retained on a quasi-government basis in order to provide information about the coal industry generally. Finally, Hobart House could either be sold, sub-let or converted into an hotel in order to generate yet more income for the company.

Chapter
16

A price for the industry

WHAT IS IT WORTH?

An asset is only worth what someone else will pay for it and this is as true of British Coal as it is of any other sector of industry. The price an investor is willing to put on an asset is related to what that asset will earn and whether or not there are any other potential purchasers who would force up the price through competition. That there is competition for British Coal's assets is undisputed, with giant companies such as Hanson and much smaller ones like the fledgling consortium put together by the Union of Democratic Mineworkers and Lloyds Merchant Bank, all expressing interest. As a result it is likely that British Coal's earnings potential will be the most accurate guide to the price the government will attain for the industry when it is privatized.

First it is necessary to determine exactly what British Coal owns and then to determine how these interests influence the value of the company as a whole. As British Coal has been so streamlined in recent years there is little left which is not a major constituent part of the British coal mining industry and which therefore needs to be valued separately. The only major interests in this category are some of the freehold land and buildings (where not specifically associated with mining) and the stake in Coal Developments

(Queensland). The value of these constituent parts can be estimated and then consolidated to produce a value for the whole company.

COAL DEVELOPMENTS (QUEENSLAND)

Probably the most controversial of British Coal's non-UK mining interests is its investment in Coal Developments (Queensland) Limited. This company is a holding company for Coal Developments (Germans Creek) Pty Ltd which, in turn, owns 12.06% of the Capricorn Coal Development joint venture in Queensland, Australia. Coal Developments (Queensland) is 89% owned by British Coal, with the remainder of the ordinary shares held by Commercial Union Assurance. The shareholdings are shown in Table 16.1.

The company was set up in the 1970s to produce coking coal in Queensland and production began in 1982. Although the operation made a trading profit in the financial year ending March 1991, the need to meet interest payments meant that the company was forced into loss at the bottom line. This situation has existed since the company was formed and the profit and loss account showed a carried forward loss of £23.1 million in March 1991.

One reason for the loss was the depreciation which was charged on the company's assets as, without this, the operation would have reported a profit for the financial year. However, this charge is necessary as it represents the annual cost of using those assets and shows whether or not the shareholders would ever be likely to generate a positive return on their initial investment. This currently looks unlikely as although the operation is generating cash the earnings figures are not calculated on a commercial basis. This is because

Table 16.1 Shareholdings in Coal Developments (Queensland)

Class	Holder	Number	Par value	Share capital (£)
'A' Ordinary	British Coal Corporation	75 390	5p	3 769.50
'B' Ordinary	Commercial Union Assurance	9 030	5p	451.50
				4 221.00
Deferred	Shell Co. of Australia	8 879	£1	8 879.00
			Total	13 100.00

Table 16.2 Loans to Coal Developments (Queensland) (£000)

	1990	1991
British Coal	9 936.9	9 959.5
Commercial Union Assurance	279.4	283.7
Total	10 216.3	10 243.2

Coal Developments has raised loans from both British Coal and Commercial Union which do not bear interest. The breakdown of these loans in 1990 and 1991 is shown in Table 16.2; if they had borne a commercial rate of interest then the company would have suffered a loss and a net outflow of funds even after adding back the £1.6 million depreciation charge. In addition to these loans, Coal Developments (Queensland) has issued a series of Euro Commercial Paper notes, amounting to £22.4 million in 1991. These notes are secured by guarantee by British Coal and, therefore, effectively count as British Coal's own borrowings.

Unfortunately, even if the omission of a full interest charge enables the company to report a profit in the future it is unlikely that it will ever be able to cover all of the losses which have been sustained in the past. British Coal will, therefore, have lost taxpayers' money through its Australian adventure. This situation may be altered if the joint venture is successful in its exploration and evaluation of a nearby steam coal resource but then British Coal would be producing steam coal in Australia in direct competition to its own mines in Britain and the excuse that the coking coal did not compete with any UK-produced coals would not be credible.

Reported earnings for Coal Developments (Queensland) in 1990 and 1991 are shown in Table 16.3 and it should be noted that the accounts have consistently been delivered to Companies House much later than allowed by normal regulations. As a result the company has had to apply to the Secretary of State at the Department of Trade and Industry for special dispensation to deliver the accounts two months late in 1991 and three months late in 1990. If the accounts had been drawn up for British Coal's annual results for the year to end March they should surely have been ready to be delivered to Companies House on time!

The value of Coal Developments (Queensland) based on its earnings history and on the debt which would have to be taken on by any purchaser would probably be negligible at present. However, it would be worthwhile to British Coal to sell the holding if only for a nominal figure of £1 as it would absolve itself of the guarantee over the borrowings and obtain repayment of the

Table 16.3 Reported earnings for Coal Developments (Queensland), 1990 and 1991 (year end December, £000)

	1990	1991
Turnover	10 523.1	12 490.6
Cost of sales	8 065.0	8 714.5
Depreciation	1 557.7	1 635.0
Gross Profit	900.4	2 141.1
Distribution costs	1 018.1	1 226.1
Administration	95.0	119.5
Trading profit/(loss)	−212.7	795.5
Profit/(loss) on sale of fixed assets	151.5	−67.8
Interest income	110.8	351.6
Interest paid	1 795.8	1 593.7
Loss for the year	−1 746.2	−514.4

nearly £10 million owed. The most likely purchaser would be the other member of the joint venture which may wish to increase its interest in the operation.

COAL PRODUCTS

Coal Products is by far the largest and least contro-versial of British Coal's non-mining operations. It is a wholly owned subsidiary which specializes in the manufacture of coke and smokeless fuels, and associ-ated by-products including tar and benzole. The majority of Coal Products' raw material is coal and it provides the company with an important down-stream diversification producing an added value product from the basic coal feedstock. Because the company is dependent on coal as a raw material it is likely that Coal Products will benefit from the reduction in price which British Coal is offering the electricity generators. Certainly if there is no reduction in price then the low level of profitability would indicate that British Coal could find better uses for its capital than keep the company in operation.

Although it is unlikely that Coal Products would be able to retain all of the reduction in the price of its raw materials it can be expected that not all of the reduction will be passed on to its customers. Additionally the company has

Table 16.4 Actual and predicted earnings of Coal Products to 1995 (year end March, £m)

	1990A	1991A	1992*	1993E	1994E	1995E
Turnover:						
Coke	170.6	153.8	145.0	147.9	150.9	153.9
Chemicals	18.2	12.3	12.0	12.2	12.5	12.7
Intra group sales	−16.2	−1.6	0.0	0.0	0.0	0.0
Total	172.6	164.5	157.0	160.1	163.3	166.6
Operating costs:						
Raw materials	118.5	98.7	91.0	93.7	79.1	79.8
Other operating	57.9	59.7	60.0	62.4	64.9	67.5
Depreciation	4.7	4.0	4.0	4.2	4.3	4.5
Total	181.1	162.4	155.0	160.3	148.3	151.8
Operating profit	−8.5	2.1	2.0	−0.1	15.0	14.8

A = actual. E = estimate.
* Total turnover and operating profit figures are actual, others are estimates.

recently completed a refurbishment programme and this should also help to increase earnings in the future. The forecast turnover has therefore been increased at a rate of 2% a year over the next five years (the period over which coal prices are falling in real terms) rather than the 4% annual inflation rate which has been assumed in all other calculations. With this increase in revenues, and the anticipated decline in raw material costs, the earnings of the company are likely to improve considerably, as shown in Table 16.4.

When consolidating these earnings there are various other amounts which have to be taken into account and they are therefore not exactly the same as the figures shown in British Coal's own report and accounts. Nevertheless, it is expected that the value of Coal Products and its subsidiaries would be in the region of £100 million to £120 million. This figure is substantially less than the £489 million Anglo United paid for Coalite in 1990. One reason for this is that Coalite had a much larger spread of activities (it even included the Falkland Islands Company) which, together with its higher level of profitability, enhanced its value. Nevertheless, in the context of current market conditions Anglo United overpaid for Coalite as shown by the company's need to enter into a financial restructuring package with its bankers.

OTHER SUBSIDIARIES/ASSOCIATES

Of the other subsidiaries listed in British Coal's annual report only National Fuel Distributors is of any significance. This company, as its name suggests, organizes the transport of coal across the country. British Coal has also recently bought the 50% of British Fuels Group which it did not previously own and has now merged the two companies together. This distribution operation is a necessary part of the main mining operation and its value will be appreciated in the valuation of the coal mining operations below.

Coal Industry Estates was formerly a land and estate management company. The company is now essentially dormant with all of the farms and houses sold off; it is owed some £2.7 million by British Coal but does not earn any interest on this loan and may be wound up at some stage in the future.

British Coal International is a company limited by guarantee which was set up to promote the sale of coal mining and associated technology overseas. In its annual report for the year to end March 1992 the company states, prophetically, 'the low cost of coal on the international markets, the worldwide recession, cheap freight costs and increasingly fierce competition for overseas markets continue to have an adverse effect on overall business'.

British Coal also has interests in mining consultancy and computer services firms which have evolved from its direct interests in, and need for, these services. Additionally, Burnpark (a boiler installation firm) is a subsidiary whilst methane gas and other carbon products are produced by associates both in the UK and the USA.

BRITISH COAL MINING OPERATIONS

The size of the industry and the large number of changes across it mean that it is not possible to calculate the earnings of British Coal's indigenous coal mining operations on a pit by pit basis. The Corporation's future earnings potential is further clouded because it remains unclear which pits will still be open in one, let alone five years' time. As a result it is necessary to construct an average picture of the industry and this has been achieved using the costs of production at different levels of output anticipated by British Coal. Tables 16.5–16.8 show the estimated level of output, costs and profitability of the British Coal mining operations, together with the earnings from the subsidiaries and associates over the next eight years together with two years' historic data. As can be seen, the figures do not take into account any additional savings which could be wrought if the industry was transferred into the private sector, such as a reduction in the size of the head office and quasi-government staff at the Corporation's headquarters in London and at the various regional offices (although they do anticipate a halving of overheads from 7p/GJ to around 3.5p/GJ in 1993/4).

The tables also show that increasing the total amount of production increases the average cost of production because it means that a larger number of higher cost mines have to be kept in operation, rather than it being possible to achieve the increases in output by increasing production at the lower cost mines. For example, increasing output to 50 mt in 1997/8 would increase average costs of production to £1.30/GJ but would mean that the top 5 mt of output would be produced for a loss of £40 million. It is also the case that the figures are calculated on the basis of the current working practices legislation, rather than on the possibly more lenient restrictions on the length of time that miners are allowed to work underground which are expected to be introduced in the future.

The final figure in the tables is a net present value (NPV) which discounts the future earnings of the operation back to a 1992/3 value. This discounting is carried out on the basis of the opportunity cost of the investment rather than through an attempt to predict future inflation rates to produce a real present value of the future earnings. This is necessary because it shows the maximum amount a potential investor could borrow at 8% and still expect to get back all of the capital plus the interest which would have accrued over the period. What it does not do is provide any scope for profit over and above the return of capital and interest and this could be obtained either by offering a lower figure or by increasing the profitability of the operations above that envisaged in the calculations. The NPVs generated should therefore be assumed to be a full price for the industry. They have been calculated over a 10 year period from 1993/4 as this would be the likely length of time over which any borrowings could be raised and, in any case, the discounted value of earnings further into the future tends to become insignificant.

It should be appreciated that no exceptional or extraordinary items, to take account of closure and redundancy costs, are included for the 1992/3 financial year as it is currently impossible to estimate the scale of their impact on the bottom line. However, they are likely to be significant. Finally, as the costs of production of either an additional 5 mt or 10 mt from 1996/7 onwards are identical the greater 10 mt of output provides a higher earnings return from this year onwards, and therefore a higher NPV.

Table 16.5 Estimated level of output, costs and profitability of British Coal mining operations to 2000, base case (year end March)

	1991	1992	1993	1994	1995	1996	1997	1998	1999	2000
Deep mined	71.7	70.6	59	32	28	28	27	27	27	27
Open cast	17.0	16.7	16	15	14	14	14	12	12	12
Stocks	0.6	−2.3	1	3	−3	−3	−2	0	0	0
Total	89.3	85.0	76	50	39	39	39	39	39	39
Deep mined										
Number of pits	65	50	19	19	16	16	15	15	15	15
Manpower (000s)	57.3	43.7	38.0	15.0	11.5	11.5	11.0	11.0	11.0	11.0
Cash cost (£/GJ)	1.63	1.62	1.64	1.40	1.32	1.28	1.27	1.25	1.25	1.25
Cost @ 4% inflation	1.63	1.62	1.64	1.46	1.43	1.44	1.49	1.52	1.58	1.64
1993 prices (£/GJ)	1.76	1.81	1.86	1.51	1.47	1.42	1.38	1.33	1.33	1.33
£/GJ @ 4% inflation	1.76	1.81	1.86	1.57	1.58	1.60	1.61	1.62	1.68	1.75
Depreciation (£/GJ)	0.10	0.09	0.10	0.10	0.10	0.11	0.11	0.12	0.12	0.13
Open cast										
Cash cost (£/GJ)	1.33	1.39	1.40	1.35	1.30	1.25	1.20	1.15	1.15	1.15
Cost @ 4% inflation	1.33	1.39	1.40	1.40	1.41	1.41	1.40	1.40	1.46	1.51
1993 prices (£/GJ)	1.70	1.82	1.86	1.51	1.47	1.42	1.38	1.33	1.33	1.33
£/GJ @ 4% inflation	1.70	1.82	1.86	1.57	1.58	1.60	1.61	1.62	1.68	1.75
Earnings (£m)										
Revenues:										
Deep	2996	3035	2612	1196	1056	1064	1034	1040	1081	1125
Open cast	689	722	708	561	528	532	536	462	481	500
Total	3685	3757	3320	1757	1584	1597	1570	1502	1562	1625

Operating costs:

Deep	2780	2726	2303	1109	951	959	955	977	1016	1057
Open cast	539	551	533	501	469	469	468	400	416	432
Depreciation	173	156	136	76	70	72	73	75	78	82
Other (gains)/losses	6	-3	0	0	0	0	0	0	0	0
Total	3498	3430	2972	1687	1489	1500	1495	1452	1510	1571
Mining pre-tax	187	327	349	70	94	96	75	50	52	54
Coal Products	5	2	0	15	15	15	18	21	22	23
Properties	35	26	27	28	29	30	32	33	34	36
Distribution	4	4	4	5	5	5	6	6	6	6
Computers	1	1	1	1	1	1	1	1	1	1
Associates	1	0	0	0	0	0	0	0	0	0
Other	5	1	0	0	0	0	0	0	0	0
Total	51	34	32	49	50	52	56	61	64	66
Operating profit	238	361	381	119	145	148	131	111	115	120
Interest	143	93	96	61	52	37	20	4	-12	-31
Exceptional items	-17	-98	0	0	0	0	0	0	0	0
Pre-tax	78	170	285	58	93	111	110	107	128	151
Tax	0	0	0	0	33	39	39	37	45	53
Net profit	78	170	285	58	60	72	72	70	83	98
NPV			576							

215

Table 16.6 Estimated level of output, costs and profitability of British Coal mining operations to 2000, extra 5 mt sold (year end March)

	1991	1992	1993	1994	1995	1996	1997	1998	1999	2000
Deep mined	71.7	70.6	59	37	34	34	33	32	32	32
Open cast	17.0	16.7	16	15	14	14	14	12	12	12
Stocks	0.6	-2.3	1	4	-3	-3	-2	1	0	0
Total	89.3	85.0	76	56	45	45	45	45	44	44
Deep mined										
Number of pits	65	50	24	24	20	20	19	18	18	18
Manpower (000s)	57.3	43.7	38.0	18.0	14.5	14.5	13.5	13.0	13.0	13.0
Cash cost (£/GJ)	1.63	1.62	1.64	1.42	1.34	1.29	1.28	1.26	1.26	1.26
Cost @ 4% inflation	1.63	1.62	1.64	1.48	1.45	1.45	1.50	1.53	1.59	1.66
1993 prices (£/GJ)	1.76	1.81	1.86	1.51	1.47	1.42	1.38	1.33	1.33	1.33
£/GJ @ 4% inflation	1.76	1.81	1.86	1.57	1.58	1.60	1.61	1.62	1.68	1.75
Depreciation (£/GJ)	0.10	0.09	0.10	0.10	0.10	0.11	0.11	0.12	0.12	0.13
Open cast										
Cash cost (£/GJ)	1.33	1.39	1.40	1.35	1.30	1.25	1.20	1.15	1.15	1.15
Cost @ 4% inflation	1.33	1.39	1.40	1.40	1.41	1.41	1.40	1.40	1.46	1.51
1993 prices (£/GJ)	1.70	1.82	1.86	1.51	1.47	1.42	1.38	1.33	1.33	1.33
£/GJ @ 4% inflation	1.70	1.82	1.86	1.57	1.58	1.60	1.61	1.62	1.68	1.75
Earnings (£m)										
Revenues:										
Deep	2996	3035	2612	1383	1282	1293	1263	1232	1282	1333
Open cast	689	722	708	561	528	532	536	462	481	500
Total	3685	3757	3320	1944	1810	1825	1799	1695	1762	1833

Operating costs:										
Deep	2780	2726	2303	1300	1173	1174	1176	1168	1214	1263
Open cast	539	551	533	501	469	469	468	400	416	432
Depreciation	173	156	136	88	84	88	89	89	93	97
Other (gains)/losses	6	−3	0	0	0	0	0	0	0	0
Total	3498	3430	2972	1890	1726	1731	1733	1657	1723	1792
Mining pre-tax	187	327	349	53	84	94	67	38	39	41
Coal Products	5	2	0	15	15	15	18	21	22	23
Properties	35	26	27	28	29	30	32	33	34	36
Distribution	4	4	4	5	5	5	6	6	6	6
Computers	1	1	1	1	1	1	1	1	1	1
Associates	1	0	0	0	0	0	0	0	0	0
Other	5	1	0	0	0	0	0	0	0	0
Total	51	34	32	49	50	52	56	61	64	66
Operating profit	238	361	381	103	135	146	123	99	103	107
Interest	143	93	96	61	52	38	20	3	−14	−33
Exceptional items	−17	−98	0	0	0	0	0	0	0	0
Pre-tax	78	170	285	42	83	108	103	97	117	140
Tax	0	0	0	0	29	38	36	34	41	49
Net profit	78	170	285	42	54	70	67	63	76	91
NPV			524							

Table 16.7 Estimated level of output, costs and profitability of British Coal mining operations to 2000, extra 10 mt sold (year end March)

	1991	1992	1993	1994	1995	1996	1997	1998	1999	2000
Deep mined	71.7	70.6	59	41	40	40	38	38	38	38
Open cast	17.0	16.7	16	15	14	14	14	12	12	12
Stocks	0.6	-2.3	1	6	-3	-3	-1	1	0	0
Total	89.3	85.0	76	62	51	51	51	51	50	50
Deep mined										
Number of pits	65	50	28	28	25	25	23	22	22	22
Manpower (000s)	57.3	43.7	38.0	20.5	17.5	17.5	16.5	15.5	15.5	15.5
Cash cost (£/GJ)	1.63	1.62	1.64	1.42	1.35	1.30	1.28	1.26	1.26	1.26
Cost @ 4% inflation	1.63	1.62	1.64	1.48	1.46	1.46	1.50	1.53	1.59	1.66
1993 prices (£/GJ)	1.76	1.81	1.86	1.51	1.47	1.42	1.38	1.33	1.33	1.33
£/GJ @ 4% inflation	1.76	1.81	1.86	1.57	1.58	1.60	1.61	1.62	1.68	1.75
Depreciation (£/GJ)	0.10	0.09	0.10	0.10	0.10	0.11	0.11	0.12	0.12	0.13
Open cast										
Cash cost (£/GJ)	1.33	1.39	1.40	1.35	1.30	1.25	1.20	1.15	1.15	1.15
Cost @ 4% inflation	1.33	1.39	1.40	1.40	1.41	1.41	1.40	1.40	1.46	1.51
1993 prices (£/GJ)	1.70	1.82	1.86	1.51	1.47	1.42	1.38	1.33	1.33	1.33
£/GJ @ 4% inflation	1.70	1.82	1.86	1.57	1.58	1.60	1.61	1.62	1.68	1.75
Earnings (£m)										
Revenues:										
Deep	2996	3035	2612	1532	1508	1521	1455	1463	1522	1583
Open cast	689	722	708	561	528	532	536	462	481	500
Total	3685	3757	3320	2093	2036	2053	1991	1926	2003	2083

Operating costs:										
Deep	2780	2726	2303	1441	1390	1392	1354	1386	1442	1500
Open cast	539	551	533	501	469	469	468	400	416	432
Depreciation	173	156	136	98	99	103	102	106	110	115
Other (gains)/losses	6	-3	0	0	0	0	0	0	0	0
Total	3498	3430	2972	2040	1958	1964	1924	1892	1968	2047
Mining pre-tax	187	327	349	53	79	89	67	33	35	36
Coal Products	5	2	0	15	15	15	18	21	22	23
Properties	35	26	27	28	29	30	32	33	34	36
Distribution	4	4	4	5	5	5	6	6	6	6
Computers	1	1	1	1	1	1	1	1	1	1
Associates	1	0	0	0	0	0	0	0	0	0
Other	5	1	0	0	0	0	0	0	0	0
Total	51	34	32	49	50	52	56	61	64	66
Operating profit	238	361	381	102	129	140	123	95	98	102
Interest	143	93	96	61	51	36	17	-1	-19	-40
Exceptional items	-17	-98	0	0	0	0	0	0	0	0
Pre-tax	78	170	285	41	78	104	106	96	118	142
Tax	0	0	0	0	27	37	37	34	41	50
Net profit	78	170	285	41	50	68	69	62	76	92
NPV			526							

Table 16.8 Estimated level of output, costs and profitability of British Coal mining operations to 2000, extra 15 mt sold (year end March)

	1991	1992	1993	1994	1995	1996	1997	1998	1999	2000
Deep mined	71.7	70.6	59	47	45	45	43	43	43	43
Open cast	17.0	16.7	16	15	14	14	14	12	12	12
Stocks	0.6	-2.3	1	6	-2	-2	0	2	0	0
Total	89.3	85.0	76	68	57	57	57	57	55	55
Deep mined										
Number of pits	65	50	32	32	28	28	26	25	25	25
Manpower (000s)	57.3	43.7	38.0	23.0	19.5	19.0	18.0	17.0	17.0	17.0
Cash cost (£/GJ)	1.63	1.62	1.64	1.43	1.35	1.31	1.29	1.27	1.27	1.27
Cost @ 4% inflation	1.63	1.62	1.64	1.49	1.46	1.47	1.51	1.55	1.61	1.67
1993 prices (£/GJ)	1.76	1.81	1.86	1.51	1.47	1.42	1.38	1.33	1.33	1.33
£/GJ @ 4% inflation	1.76	1.81	1.86	1.57	1.58	1.60	1.61	1.62	1.68	1.75
Depreciation (£/GJ)	0.10	0.09	0.10	0.10	0.10	0.11	0.11	0.12	0.12	0.13
Open cast										
Cash cost (£/GJ)	1.33	1.39	1.40	1.35	1.30	1.25	1.20	1.15	1.15	1.15
Cost @ 4% inflation	1.33	1.39	1.40	1.40	1.41	1.41	1.40	1.40	1.46	1.51
1993 prices (£/GJ)	1.70	1.82	1.86	1.51	1.47	1.42	1.38	1.33	1.33	1.33
£/GJ @ 4% inflation	1.70	1.82	1.86	1.57	1.58	1.60	1.61	1.62	1.68	1.75
Earnings (£m)										
Revenues:										
Deep	2996	3035	2612	1757	1697	1711	1646	1656	1722	1791
Open cast	689	722	708	561	528	532	536	462	481	500
Total	3685	3757	3320	2317	2225	2243	2182	2118	2203	2291

Operating costs:										
Deep	2780	2726	2303	1664	1564	1578	1544	1581	1645	1710
Open cast	539	551	533	501	469	469	468	400	416	432
Depreciation	173	156	136	112	112	116	116	120	125	130
Other (gains)/losses	6	−3	0	0	0	0	0	0	0	0
Total	3498	3430	2972	2277	2144	2163	2128	2101	2185	2273
Mining pre-tax	187	327	349	40	81	80	54	17	18	18
Coal Products	5	2	0	15	15	15	18	21	22	23
Properties	35	26	27	28	29	30	32	33	34	36
Distribution	4	4	4	5	5	5	6	6	6	6
Computers	1	1	1	1	1	1	1	1	1	1
Associates	1	0	0	0	0	0	0	0	0	0
Other	5	1	0	0	0	0	0	0	0	0
Total	51	34	32	49	50	52	56	61	64	66
Operating profit	238	361	381	89	131	132	111	78	82	85
Interest	143	93	96	61	51	34	15	−3	−20	−40
Exceptional items	−17	−98	0	0	0	0	0	0	0	0
Pre-tax	78	170	285	28	80	97	95	82	102	125
Tax	0	0	0	0	28	34	33	29	36	44
Net profit	78	170	285	28	52	63	62	53	66	81
NPV			469							

OTHER COAL DEALS

Over recent years there have been many deals in the coal mining industry, some of which have taken place in the UK. However, because of the restrictions on output and on the number of people who can be employed underground there have been relatively few deals of any consequence in this country. It is therefore necessary to look overseas to discover the deals which have been struck and to attempt to determine a price that a purchaser may offer for British Coal either as a whole unit or as a composite of many separate parts. Obviously the characteristics of any deal in the UK would be different from those in Australia because of that country's large export market. In the USA, however, the situation may be more similar because of the large indigenous coal burn by electricity utilities. Coal deals where details have been released are shown in Table 16.9, together with relevant statistics.

As can be seen from Table 16.9 there is a wide disparity in the amount which companies around the world have paid for coal-producing assets. This is in terms of the amount paid both per tonne of annual coal production and per tonne of reserves. These statistics are obviously influenced by the operating profit per tonne which can be generated from the mines and this is why the price earnings ratio (PER) is important as it gives an indication of the amount which the company can expect to earn on its investment. If the mine has been suffering from poor investment and a resulting low level of profitability it may wish to pay a high PER, invest money and hope to earn a much better return. In such an instance the cost per tonne of reserves will be relatively low as the potential of earning any money from them may be minimal. Such was the situation for Elders Resources when it purchased the Saxonvale mine from BHP. Elders, and subsequently Oakbridge, invested a large amount of money in the project and increased output and profitability substantially. As a result the value of Oakbridge to McIlwraith McEacharn was over four times the value of the acquisition of Saxonvale, when calculated in terms of the cost per annual tonne of output and increased further when Ban Pu Coal bought its 5% holding.

A FINAL VALUE

Because British Coal is potentially in a similar situation to Saxonvale, as it requires the investment of time and money to increase productivity, it can be expected that the value of the Corporation to potential

Table 16.9 Deals in the US coal industry, 1987–92

Date	Companies/Deals	Value of company indicated by deal	Per annual tonne	Per reserve tonne	PER
October 1992	Peabody buys Costain Coal	US$200m	N/A	N/A	N/A
June 1992	Peabody (Martinka) buys Southern Ohio	US$165m	US$66	US$2.75	N/A
May 1992	Peabody buys Carbonar	US$6.6m	US$33	US$1.32	N/A
April 1992	Ashland buys Dal Tex Coal	US$250m	US$52	US$1.25	N/A
March 1992	Voest Alpine buys 40% of Semirara Coal of Manilla	US$45.4m	N/A	N/A	15.1
February 1992	Budge Mining management buyout	£106.5m	N/A	N/A	N/A
January 1992	Nerco sells 2 more mines	US$7.0m	US$6.4	N/A	2.3[1]
December 1991	Nerco sells 2 mines	US$177m	US$22.95	US$0.94	44.3
November 1991	RWE buys 50% of Du Pont Coal	US$1780m	US$32.96	N/A	8.7
March 1991	CRA bids for Coal and Allied Industries	US$370m	US$35.35	US$0.24	8.0
February 1991	Amax buys Cannelton Coal	US$100m	US$18.37	US$0.80	N/A
January 1991	Arco Australia increases Gordonstone stake	US$392m[2]	US$93.33	N/A	N/A
January 1991	Ban Pu Coal buys 5% of Oakbridge	US$280m	US$38.24	US$0.17	12.1
August 1990	McIlwraith McEacharn buys Oakbridge	US$216m	US$29.50	US$0.13	9.3[3]
July 1990	Queensland Coal Trust buys South Blackwater Mine	US$117m	US$58.50	N/A	N/A
July 1990	Hanson buys Peabody	US$1320m	US$16.10	N/A	25.7
January 1990	Ashland buys Mingo Logan	US$144m	US$79.77	US$1.23	N/A
December 1989	Rand Mines bids for Middleburg Colliery	US$139m	US$25.27	US$1.68	N/A
August 1989	Newmont lifts Peabody stake	US$1150	US$14.02	N/A	18.2
January 1989	Ryan International management buyout	£69.6m	£11.60	£2.32	11.4
November 1988	Geevor buys Mainband	£5.0m	£50	£1.67	N/A
August 1988	Elders Resources buys troubled Saxonvale Mine	US$56.8m	US$6.76	US$0.06	N/A
June 1988	Young Group Flotation	£13m	£59.09	N/A	N/A
January 1987	Floyd Oil buys Hampton Coal	£5.75m	N/A	N/A	15.0

[1] Based on operating earnings.   [2] Project cost.   [3] Abnormal items excluded from calculation.

purchasers will reflect this. Therefore, although there may be a relatively low value in terms of the cost per annual tonne and cost per reserve tonne there is likely to be a relatively high PER put on the operations. Most of the ratings for ongoing and profitable mining operations have been in the region of 8 to 10 times annual net earnings. Some of the deals, however, have been completed on higher ratings of 15 to 25 times and this is expected to be the likely range of historic sustainable earnings required to pay for British Coal. As there is now only a reduced scope for future improvement in productivity, as there has been such a significant increase over the past few years (see Figure 11.1) it is likely that the price will be towards the bottom of this range. This should be especially because the mines are likely to have relatively short term sales contracts, which reduce the longer term security of an operation as it may have to compete in the open market for a larger proportion of its sales (note the relatively high price paid by Peabody for Southern Ohio (Table 16.9) which has a 20 year supply contract). A rating of some 12 times projected 1993/4 earnings would therefore be appropriate for British Coal.

In Table 16.9 the earnings used in calculating the PERs have been historic figures. This will not be appropriate for British Coal because of the massive fall in forecast output whether or not there is some sort of compromise deal worked out in the meantime. As a result the earnings figures used for calculating the PER should be those of the financial year 1993/4 when the company expects to sell a minimum of 40 mt to the electricity industry. These figures, extracted from Tables 16.5–16.8, are shown in Table 16.10, together with the value this places on British Coal as a whole.

Assuming that the mines have around 40 years of reserves at the more stable 1994/5 production levels, then the cost per annual 1993/4 tonne and the cost per reserve tonne on these figures are as shown in Table 16.11. As can be seen, these figures are broadly in line with the figures in Table 16.9. It

Table 16.10 The value of British Coal, based on earnings figures for 1993/4

	NPV		93/4 earnings	Value @ 12 times PER	
	(£m)	(US$m)	(£m)	(£m)	(US$m)
Base case	576	864	58	696	1044
+5 mt	524	786	42	504	756
+10 mt	526	789	41	492	738
+15 mt	469	704	28	336	504

Note: US$ figures converted at US$1.5 = £1.

Table 16.11 The cost per annual 1993/4 tonne and the cost per reserve tonne on the figures in Table 16.10

	NPV values				*PER values*			
	Per annual tonne		*Per reserve tonne*		*Per annual tonne*		*Per reserve tonne*	
	(£)	*(US$)*	*(£)*	*(US$)*	*(£)*	*(US$)*	*(£)*	*(US$)*
Base case	11.52	17.28	0.37	0.56	13.92	20.88	0.45	0.68
+5 mt	9.36	14.04	0.29	0.44	9.00	13.50	0.28	0.42
+10 mt	8.48	12.72	0.26	0.39	7.94	11.91	0.24	0.36
+15 mt	6.90	10.35	0.21	0.32	4.94	7.41	0.15	0.23

Note: US$ figures converted at US$1.5 = £1.

can therefore be assumed that the price that British Coal mining operations will fetch will be in the region of £500 million to £600 million depending on the changes to the size of the industry over the next few years. In addition to this, the value of the freehold land and property not included with the coal mining assets (particularly Hobart House) has to be considered. British Coal has not revalued its assets and it is therefore difficult to ascertain the true value of these holdings – especially to a potential developer. Nevertheless, in their review of the organizational efficiency of British Coal, Ernst & Young state that British Coal estimate the non-operational freehold property interests to be worth between £300 million and £1000 million depending upon whether or not they were the subject of a forced sale or an orderly disposal. It should, finally, be noted that this means that the Corporation is worth significantly less than the total amount of redundancy and other subsidies (which are in excess of £18 billion) that the government has pumped into the industry over the past 14 years. If anything this must be a stark lesson of the errors of government intervention in industry.

Chapter

17

The final countdown

THE CONTRACT DEBATE

In early 1992 British Coal started formal discussions with the electricity generators, National Power and PowerGen, over the new coal supply contracts that it hoped would secure the future of the industry. These discussions proved fruitless due to the problems of overcapacity in the electricity generation industry and the relatively high price of British Coal's product. This meant that if they bought their supplies from British Coal the generators would be producing electricity at a higher price than could be achieved by purchasing coal on the free market. As a result the generators refused to sign any contracts unless they were guaranteed a market for the electricity produced through back-to-back contracts with the electricity distributors – the RECs. As far as the RECs were concerned they were unwilling to purchase the electricity until they had been given an assurance by Offer that they would not be in breach of their obligations under the 1989 Electricity Act to purchase electricity as cheaply as possible.

The RECs were also faced with a second problem. This covered the franchise market which each REC had as a captive market within its geographical area of operation. Following the privatization of the electricity supply industry the franchise market was limited to all consumers using less than 1 MW of electricity but, as part of the process of opening the electricity market up to more competition, the restricted franchise market is to be reduced to all consumers using less than 100 kW in April 1994 and will be abolished altogether from April 1998. At present there are some 4000 customers which require a supply of over 1 MW, whilst the reduction in the size of the franchise market to 100 kW will increase this figure by some 40 000 additional consumers. The RECs were painfully aware that they would only be able to pass on the higher cost of coal generated electricity to their captive franchise market customers and the decline in the size of this market is the reason for the reduction in the coal purchase contracts from 40 mt to 30 mt between 1993/4 and 1994/5.

As the negotiations progressed British Coal became aware that it had vast operational overcapacity and this was the reason for the massive cut-backs announced on 13 October 1992. Indeed, when the discussions started British Coal had hoped to be able to negotiate contracts to supply 50 mt reducing to 40 mt rather than the 40 mt and 30 mt that have now been agreed. The Coal White Paper which was published on Thursday 25 March 1993 stated that 'it was not until 12 October that British Coal informed the President of the Board of Trade of the number and names of the pits it proposed to close'. However, the DTI must have been aware of the impending scale of the cut-backs necessary – especially following the Rothschild report which suggested that only 14 pits would survive privatization – and to use the lack of knowledge of whether 31 or 32 pits would have to close is no excuse for its mishandling of the situation. Also it must be noted that the closure of three of the pits had previously been announced. It remains unclear why British Coal felt obliged to announce their closure a second time, thereby boosting the total number of closure pits which it released to the public.

A NATION SHOCKED

The public outcry that followed the pit closure announcement, coming so hard on the heels of the currency fiasco of Black Wednesday, sent the government into a spin and further wrecked confidence in the Conservative administration. Indeed, the *Newsnight* presenter, Jeremy Paxman, summed up the nation's feelings when he stated on 14 October that 'where the government ought to be showing leadership it is displaying impotence'. In one respect the government was fortunate that Parliament had not reconvened following the summer recess as the initial scenes from the

House of Commons would have shown it in complete disarray. Luckily, however, they were able to consider how best to handle the potential back-bench revolt over the closures over a longer period of time and this may have saved the Tories from defeat on the pit closure debate on Wednesday 21 October.

When Parliament reconvened on Monday 19 October Michael Heseltine made up for his inability to advise the House of Commons of the impending pit closure programme by making a statement to MPs on the decision and the reasons behind it. He also agreed that it would have been better to have made the statement first to Parliament but said that the escalation of rumours surrounding the impending announcement meant that it had to be brought forward. As a result he said 'I regret this discourtesy to the House'. In another comment he stated that 'I accept full responsibility for this decision' and it must have been this more than anything that raised the blood of some of the Labour Members of Parliament, such as Dennis Skinner the MP for Bolsover and Secretary of the Parliamentary Miners Group, sensing that there could be a political killing to be made.

The following Wednesday the formal debate on the coal industry took place. By this stage the government had climbed down on the number of pits to be closed immediately, from the initial 31 to 10, and this, together with the promise of a full review of the prospects for the coal industry, enabled the government to squeeze through with a majority of 13 votes against its 21 seat absolute majority at the General Election. One of the government's main areas of political manoeuvering was to secure the abstention of the Ulster Unionist MPs in a behind the scenes deal, otherwise the six Tories who voted against the government would have reduced its majority to only three.

The main questions raised in the debate centred on the ability of the government to ensure that the review into the pit closures would be fair. This was not just because the opposition was worried that the review would be slanted towards the government's intention to privatize the industry, but also because it would be aided by British Coal which was the originator of the closure proposals. The review was called for by the Labour Party in the run-up to the debate with John Smith, the Leader of the Opposition, asking in Prime Minister's question time on 20 October 'if the Prime Minister really believes he has a strong case what has he to lose, what is he afraid of [in an independent review]?' The fact that a review was offered to appease the Tory backbenchers led Robin Cook to comment 'how pleased they [the government] must be that we did not ask for a General Election or they might have offered that too!'

After the debate the dust started to settle in Parliament, although the mining communities remained in a state of severe shock. The decision to continue with the closure of ten pits provoked strong comments because it seemed that the pits were to be condemned even though the outcome of the review might have provided good reasons for them remaining open. It also led to various challenges by the mining unions in the High Court because British Coal had not complied with employment legislation and had not notified the employees of the impending redundancies in advance of the closures. This would have enabled the unions to put forward their arguments that the pits should remain open but also would have given the miners a greater opportunity to prepare for their loss of employment.

This High Court action continued to rumble on into 1993 and, even in April 1993 (i.e. after all of the 31 pits on the original list were due to be closed) British Coal and the unions were unable to reach an agreement on how the Colliery Review Procedure should look at the closure of the pits. British Coal wanted to employ Boyds to investigate the potential of the pits on a combined basis whilst the unions wanted the closures considered on a pit-by-pit basis. The unions also had some disquiet about the choice of Boyds as the consultants covering the closures as they were one of the consultants employed by the DTI to aid it in its review of the coal industry. Boyds had also been employed by British Coal to look at the prospects of the mines in the run-up to the pit closure announcement and the firm's ability to change its original view was questioned by the unions.

THE REVIEWS

In order to aid in its review of the prospects for the coal industry and in preparation of the White Paper the DTI appointed four firms of consultants to carry out reviews of specific areas affecting the profitable future of the coal industry. These were in addition to the other firms that had already been appointed to advise the government and British Coal in the run-up to privatization. The consultants, listed below, produced their reports on 22 January 1993, at a cost to the DTI of £2m.

1. PIMS, a management consultancy firm which concentrated on the comparison of British Coal's performance against a range of other international companies operating in similar industries.
2. Ernst & Young, the accountants and management consultants, which were employed to review the organizational efficiency and overhead structure of British Coal's mining activities, and also provided some international comparisons.

3. Caminus Energy, a specialist energy consultancy, which was asked to research the outlook for the coal market in the UK and British Coal's potential share of that market.

4. John T Boyd (Boyds), which has already been mentioned, was employed to advise on the prospects for the 21 pits on which the moratorium on closure had been placed. The company also reviewed the 19 pits due to remain in operation plus Asfordby, which is being developed, to confirm that their position on the operational list was justified.

In the end Boyds produced two reports; the first was the subject of the basic review, and the second was to check that the 10 pits on the closure list should be closed. As an adjunct to the second review Boyds was also asked to confirm that British Coal was maintaining the 10 pits in good working order so that they could be reopened if the review warranted continued operation. Disturbingly for British Coal Boyds detailed two pits, Taff Merthyr and Grimethorpe, which it stated should not have been closed as they did not meet the criteria for closure which British Coal stated that it had used when drawing up the closure list. In the case of Taff Merthyr it agreed that the low level of reserves was a mitigating factor although there were some 122 000 tonnes of reserves available for extraction at the time of the cessation of production on 13 October. Boyds stated that leaving this coal in the mine 'deprived the workforce of enhanced earnings' but because of the low level of reserves reopening the pit could not be economically justified.

At Grimethorpe, on the other hand, Boyds concluded that the pit did not meet one of the closure criteria because it was making a profit at the time of the closure announcement. Indeed, Boyds estimated that the colliery 'could reasonably be projected to achieve profitable operation' with a projected margin above overall costs of around £0.10/GJ or around £2.38/t assuming a relatively low energy content of the mine's coal. On annual production of just over 1 million tonnes Boyds' projected figures indicate that Grimethorpe could generate over £2 million of annual profit subject to a market for the mine's output being available. These figures were based on the assumption that there was no change to basic mining legislation and that existing work practices were largely unaltered.

In a second scenario, where new legislation was introduced which served to enable British Coal to increase output through the working of fewer but longer shifts it was projected that Grimethorpe would have a projected profit margin of around £0.22/GJ on increased output of 1.165 mt. Again using the base of 23.8 GJ/t Boyds' figures indicate that the mine would make annual profits of some £6.1 million. Unfortunately for Grimethorpe it is situated adjacent to another mine on the closure list, Houghton Main, and this colliery could not be seen as potentially profitable under either scenario. This was partially because of its low output of only around 400 000 t/year but also because of the relatively high costs associated with the Southside wash plant which both collieries used on a joint basis.

British Coal stated that the closure of Houghton Main colliery, which Boyds viewed as justified, resulted directly in the decision to close Grimethorpe because of links between the two collieries. Whilst the two collieries were linked underground this would not mean that Grimethorpe would have to bear higher water pumping charges because Houghton Main was a dry mine. At Grimethorpe annual pumping amounted to some 160 million gallons. The Southside coal washery plant used by both collieries was admitted by Boyds as being a problem as far as the costs of operation were concerned and the consultants suggested that the washery was closed and future coal washed on a contract basis. Boyds suggested that this would lead to a reduction in coal washing costs from some £5.86 per saleable tonne to only £3.00 per saleable tonne. As a result 'Boyds reject the British Coal assertion that Grimethorpe and Houghton Main are inextricably linked'. Therefore, the difficulty of finding a market for Grimethorpe's output was British Coal's only defence in closing the mine, and this despite its contract to supply 360 000 t of coal to ICI's Monckton plant (see Chapter 11).

Under the rigours of the free market implied by the second scenario there was only one other colliery on the list of 10 due to be closed that Boyds considered could be profitable: Markham Main. Unfortunately for Markham Main it was constrained by two factors. First, the mine needed a large amount of capital expenditure to develop a new single retreat longwall face with bolted two-entry development. Secondly, the mine required a 'substantial improvement in the work ethos before the Case II mining plan and cost projections could be achieved'. Boyds admitted that both of these assumptions involved 'a high degree of risk given the radical departure from existing mining operations' and it therefore concluded that the closure was justified in the light of developments elsewhere in the industry.

The main results of the Boyds reviews into the 20 continue to operate and 21 moratorium pits under Case I and Case II are shown in Tables 17.1 and 17.2. Whilst it may appear that the Ellington Colliery (which appears as the lowest ranked of the collieries on the list of those which are set to continue to operate) should close, it has a specific advantage due to its direct conveyor link to Alcan's Lynemouth Power Station and Aluminium Smelter which is sufficient justification for the pit to remain in operation.

Concurrent with the DTI review of the coal industry the Trade and Industry Select Committee and the Employment Select Committee decided

Table 17.1 The 20 continue to operate pits

Mine	Case I			Case II			
	Production (000t)	Margin (£/GJ)	Rank	Production (000t)	Mine life (Years)	Margin (£/GJ)	Rank
Daw Mill	1860	0.44	1	2212	23.1	0.57	1
Wistow	2285	0.42	2	2283	13.7*	0.49	2
North Selby	1762	0.28	3	1930	13.7*	0.37	5
Riccall	2068	0.26	4	2111	13.7*	0.33	8
Thoresby	1870	0.27	5	2385	10.7	0.47	3
Goldthorpe	1400	0.27	6	1400	4.3	0.42	4
Asfordby	1644	0.18	7	1930	46.2	0.31	7
Manton	1315	0.23	8	1493	20.4	0.34	9
Harworth	2047	0.22	9	2462	30.4	0.38	6
Longannet	1693	0.19	10	1774	7.4	0.26	13
Welbeck	1565	0.17	11	1793	12.9	0.31	11
Kellingley	1839	0.14	12	2223	35.1	0.32	10
Whitemoor	1213	0.14	13	1347	13.7*	0.25	12
Ollerton	1511	0.15	14	1260	14.3	0.15	17
Annesley	1320	0.12	15	1476	4.6	0.25	14
Stillingfleet	2365	0.08	16	2730	13.7*	0.20	15
Silverwood	892	0.03	17	966	29.7	0.11	18
Littleton	1038	0.03	18	1121	15.4	0.19	16
Tower	666	−0.06	19	753	4.5	0.09	19
Ellington	1086	−0.55	20	1304	11.5	−0.29	20

* Selby complex average life

that they would investigate the background to the pit closures within the remit of their specific areas of interest. Of these the Trade and Industry Committee is of most interest and its members are listed in Table 17.3. As can be seen from this list the Conservative members could exercise a majority of 6 to 5, assuming that the Chairman was allowed a casting vote.

The Committee started its deliberations on Tuesday 27 October 1992 when Michael Heseltine was called to give evidence, and completed its public hearings on Tuesday 15 December when Michael Heseltine again returned to answer questions on the progress of the DTI's coal review. In all the Trade and Industry Committee received public evidence from 31 different organizations, including the DTI and British Coal, and received written submissions from many other organizations and individuals. There was also a great deal of cross-fertilization of information between the Committee and the DTI. When the

Table 17.2 The 21 moratorium pits

Mine	Case I			Case II			
	Production (000t)	Margin (£/GJ)	Rank	Production (000t)	Mine life (Years)	Margin (£/GJ)	Rank
Maltby	1251	0.21	1	1520	36.9	0.41	1
Hatfield	783	0.15	2	953	31.5	0.35	2
Prince of Wales	1556	0.10	3	1825	10.6	0.27	3
Frickley	1046	0.10	4	1298	12.4	0.27	4
Point of Ayr	542	0.04	5	637	10.2	0.18	6
Bentley	801	0.00	6	899	7.3	0.11	9
Rossington	803	−0.02	7	1114	26.5	0.28	5
Wearmouth	1337	−0.06	8	1578	35.1	0.12	7
Calverton	905	−0.08	9	973	10.3	0.02	12
Kiveton Park	904	−0.09	10	989	6.0	−0.01	14
Bilsthorpe	1292	−0.13	11	1472	23.1	0.04	11
Silverdale	980	−0.11	12	1172	18.3	0.14	8
Shirebrook	1286	−0.13	13	1553	14.1	0.08	10
Markham	1359	−0.15	14	1420	8.8	−0.05	16
Bolsover	650	−0.23	15	650	1.8	−0.38	19
Clipstone	814	−0.26	16	931	7.0	0.01	13
Rufford	716	−0.28	17	874	14.0	−0.03	15
Sharlston	900	−0.38	18	900	2.0	−0.23	18
Bevercotes	708	−0.42	19	770	39.8	−0.19	17
Westoe	1141	−0.52	20	1401	16.4	−0.46	20
Easington	1200	−0.55	21	1200	5.6	−0.55	21

Table 17.3 Members of the Trade and Industry Select Committee

Name	Constituency	Party
Mr Richard Caborn (Chairman)	Sheffield Central	Labour
Mr John Butterfill	Bournemouth West	Conservative
Mr Malcolm Bruce	Gordon	Liberal Democrat
Mr Michael Clapham	Barnsley West and Penistone	Labour
Dr Michael Clark	Rochford	Conservative
Sir Anthony Grant	Cambridgeshire South West	Conservative
Dr Keith Hampson	Leeds North West	Conservative
Mr Adam Ingram	East Kilbride	Labour
Rt Hon Cranley Onslow	Woking	Conservative
Rt Hon Stanley Orme	Salford East	Labour
Mr Barry Porter	Wirral South	Conservative

public hearings had been completed the Committee compiled a major report on 'British Energy Policy and the Market for Coal' which contained a total of 39 recommendations for the government to adopt to reduce the adverse impact of the pit closures. In an amazing example of cross-party unity the report was unanimously adopted by the Committee, although there now appear to have been some areas where the Committee members did not actually agree on all of the report's recommendations.

Specifically, this apparent disunity became apparent during the 29 March 1993 debate on the White Paper on the future of the coal industry where Cranley Onslow mentioned that there were some 'passages in the report where the conclusions did not carry the full stamp of detailed consideration' due to the lateness of the hour and an agreement to disagree. Cranley Onslow also mentioned that the form of words used in some of the recommendations was of prime importance, specifically where the Committee suggested that the 'Government should consider' and that the 'Government should explore the possibility' in recommendations 15 and 23 respectively. These were related to the two areas of increasing coal sales to the industrial market and where a possible restriction on the use of gas generators was to be considered. Further Cranley Onslow admitted that 'there was a need to have more work done than we were able to do'.

In a damning indictment of British Coal's stagnation under government ownership Cranley Onslow continued that improved working practices 'should have been put in place many years ago [and] had they been there is no doubt (in my mind) that the markets it could command would have been much greater'. He also voiced his worries about the Corporation's existing management when he stated that 'I don't think that British Coal alone can be relied upon to do what still needs to be done'.

Barry Porter was another member of the Committee who mentioned in the White Paper debate that the Tory and Labour members sat staring at each other at 10 pm and that the figure of 19 mt of additional coal for which the Committee suggested that there was an additional market was achieved 'by good old fashioned political horse trading'. He abstained from the final vote to put the report to the House and mentioned that he thought that the figure of '12 mt was about right'. As a result of all of the deliberation the report was not finally agreed until 26 January and it was released to the public on 28 January 1993.

THE PROSPECTS FOR COAL

The White Paper was presented to Parliament on Thursday 25 March 1993 with a full debate on the proposals taking place on the following Monday. Many MPs voiced their disquiet about being given only the weekend to digest all 152 pages of the document but the fact that Parliament was due to rise for the Easter recess on Friday 2 April meant that there was insufficient space in the schedule to give the MPs further time for consideration. It was also the case that the impending deadline of 31 March 1993 for the end of British Coal's existing contracts with the generators, and the lack of new contracts for April and beyond, meant that the government had to announce its intentions in time to allow the necessary new contracts to be signed. The delays in the preparation of the White Paper also meant that it had been held up to such an extent that it was considered that it would be unfair to the miners to delay their agony any further.

As a result of the proposals in the White Paper the government decided that a subsidy of 'hundreds of millions of pounds' could be justified to enable all British deep mine coal producers to sell their output at world market levels. This, it was estimated would reduce the level of imports into the UK to the extent of some 12 mt – a figure which would allow some twelve pits to remain in operation beyond the original March 31 closure date. In order to ensure that the government did not fall foul of any anti-competition legislation it was also necessary to open the subsidy to all deep mines in the country whether or not they were in the control of British Coal. Open pit operations were considered to be profitable enough not to need a subsidy, although the royalties that they paid to British Coal were to be reduced from the level of £4.50 to £5.00 a tonne to only £2.00 a tonne.

Of the nine pits for which a market could not be found it was decided by British Coal that six should be placed on care and maintenance. This would retain them in a fit state for further production to continue should any private sector interests wish to put in offers. A seventh colliery, Maltby, was to be 'placed on development'. This meant that British Coal was to spend an additional £7 million to bring the mine up to world class standards which would enable it to be competitive and be brought back into operation later in the decade when the depletion of the reserves at other mines forced their closure. Indeed, it would have been a distinct crime to close the mine for good in view of the huge amount of money that British Coal had spent in the development of the operations over the past few years. This was estimated to be £183 million by Kevin Barron the MP for Rother Valley in whose constituency the pit lies. That British Coal recognized this can be seen by its position on the original list of four pits which were to be placed on care and maintenance under the original proposals, although it was also on the Rothschild list of fourteen pits which would still be in production following privatization.

Finally, the low standing of both Bolsover and Sharlston meant that they were condemned to closure along with the ten pits from the original 31 for which a closure moratorium was not granted. These two pits fell towards the bottom of the Boyds review of the 21 pits, Sharlston's position was at number 18 under both of Boyds scenarios, whilst Bolsover fell from 15 to 19 in Case I and Case II respectively. The main problem for both collieries, which outweighed any potential increase in productivity, was that the mines had a very low level of reserves. These were estimated to be sufficient to last only 1.8 years in the case of Bolsover and 2.0 years for Sharlston and as a result they were the mines with the lowest levels of reserves.

However, it was privatization which became the main thrust of Michael Heseltine's argument in the White Paper which stated that 'the government's energy policy . . . centres on the creation of competitive markets' and that privatization was necessary to ensure that a competitive market could be created because of the 'unwelcome legacy . . . that state-owned industries have often made uncommercial investments, or preserved activities well beyond the time when their commercial rationale ran out, often as a result of political intervention. Privatization forces such issues to be confronted and reveals the real costs of economic activities, which have often been disguised in the past'. Such, of course, was the case with the nuclear power industry.

The one area on which the White Paper touched only in the broadest terms, however, was the very subject of the privatization of British Coal. And this was despite the massive amount of time and effort which had gone into its production. The document allows for the creation of a Coal Authority which will license the country's coal mines and will be prohibited from mining. However, it does not make clear where ownership of the coal will be vested and whether or not it will be retained by British Coal. The problem regarding the costs of continuing to keep mines on care and maintenance outlined in Chapter 15 and the potential subsidy due from the royalties may be one reason behind this lack of clarity.

The only clear decision about privatization was that it will proceed but the exact details are still up for discussion between the government and other interested parties. Specifically the White Paper states that 'the full implications of the proposals in this White Paper for the structure and timing of the privatization of the coal industry require further detailed consideration'. As a result much of the White Paper is a wasted effort, although it has alleviated the government's immediate political problems. Indeed, the government's welcoming of British Coal's commitment to 'offer to the private sector pits which British Coal does not itself wish to keep in production' should not be seen to be any more than an offer which British Coal can still reject if it decides that the price offered is not high enough. As any new operator will erode British Coal's market share the value of the pits to the Corporation is likely to be substantially higher than the value placed on them by outside operators and there is therefore the potential for a serious conflict of interest on this point. In view of this the government should transfer the licensing of mines to the proposed Coal Authority with immediate effect and it should be up to the Authority to decide whether or not British Coal had been offered a fair price for any of the collieries on the closure list (see Chapter 15).

Since 1979, Conservative governments have been trying to push the screaming child of British Coal into the private sector. However, the continuation of the closure programme and the misguided commitment that there would be no compulsory redundancies maintained the Corporation as a large and unwieldy organization. The White Paper confirmed this when it stated that 'although many miners chose to leave the industry, the requirement to find jobs for those who wanted to stay left many pits overmanned. Partly as a result, the coal that they produced was unnecessarily expensive, at a time when British Coal desperately needed to reduce costs'. The Ernst & Young review also covered the subject of overmanning in British Coal's management and non-industrial activities. It highlighted the discrepancy that 'as the number of collieries has declined over the last three years, the percentage of non-industrial staff to industrial staff has increased' (see Table 17.4). And, specifically Ernst & Young stated that 'due to this [voluntary] redundancy policy . . . certain individuals are still being paid by British Coal, despite being surplus to requirements'. That such activity was carried out by British Coal under the Conservative administration is amazing as it is so reminiscent of much of the ideas of 'full employment' propounded by the far left.

In stark contrast to this overmanning the private sector was squeezed in the opposite direction because the legislation which prevented the introduction of a fully competitive market in coal production was never repealed. Moreover, it is a travesty of the government's free market ideals that it did not abolish the restriction on the numbers of men working underground in private mines when it had the chance but chose instead to raise the limit to 150 in the 1990 Coal Industry Act. Indeed, the White Paper admits that 'almost all British Coal's pits employ more than 150 people underground [and that] significantly extending the role of the private sector in the mining of coal will therefore require

Table 17.4 Industrial and non-industrial staff

Year	1989/90	1990/91	1991/92
Colliery and other industrial manpower	73 400	63 500	48 900
Non-industrial staff	11 700	10 800	9 200
Percentage	15.9	17.0	18.8

primary legislation'. The White Paper gave no clear commitment that such legislation would be introduced – again because the government wants to restrict the competition in advance of the full privatization of the coal industry. How it expects private operators to mine a large pit on the closure list which previously employed over 1000 men with only 150 is not made clear.

CONCLUSION

There is still much thought and discussion to take place before the privatization of the British coal industry can proceed. The government's strategy has often been misguided as it has prevented the implementation of policies necessary to enable an increase in productivity to the levels necessary to ensure the long life of the industry. As a result the mines were not productive enough to compete either with imported coal or with the new generation of gas-fired electricity generators.

It is also the case that the outmoded unions prevented the implementation of new work practices in the mines and that this hindered the increase in productivity, and caused more job losses in the current wave of cut-backs than might otherwise have been necessary. It is true, however, that the total number of jobs lost may have been similar due to the compulsory redundancies that would have had to occur in earlier years. Nevertheless, a higher level of productivity may have enabled an increase in the number of collieries open for longer and this would have had the effect of increasing the total number of manyears guaranteed rather than a specific number of jobs in any one year.

The problems with the NUM seem to have led the government to attempt time and again to ignore the problem of Britain's coal mines. Much should have been accomplished following the defeat of the 1984/85 miners' strike but to the shame of the government it allowed British Coal to fall back into its pre-strike uncompetitive ways. Indeed, the mistaken commitment that there would be no compulsory redundancies was one of the biggest errors of the year-long struggle which was supposed to be about the Corporation's ability to manage the industry on an economic rather than a social basis. Following the strike British Coal should have been told in no uncertain terms that it would start to receive lower prices for its output from the following year. This decline in prices should have brought the coal the Corporation sold to world levels much sooner than the 1997/8 date that is now projected. Without such a definite target to aim for the uncompetitive ways continued and the loss of electricity market share and hence the closures were inevitable.

What Michael Heseltine must now do is to capitalize on the changes that have been wrought in the industry and he must proceed with the introduction of new legislation during the rest of 1993 and privatization in 1994 at the latest. To fail to do this will leave the industry again lacking the controls of the market and will leave it open to the problems of high costs and low international competitiveness which have caused so many problems in the past. The White Paper is a pretty inauspicious start – it is to be hoped that having escaped the jaws of the political lion Tarzan soon returns to the jungle to complete his mission.

As this book goes to press British Coal has announced its plans to forge ahead with the closure of three of the 12 pits that were reprieved as a result of the White Paper. This is a direct result of the mistaken way in which the privatization process is proceeding and would not have occurred if the government/British Coal followed some of the recommendations outlined in this book. Both the government and British Coal can clearly be seen to be attempting to protect their own market at the expense of allowing other willing participants to buy into the industry. This is shown by the government's failure to remove the restrictions on other operators, which prevent them producing coal on the same terms allowed to British Coal. As a result it is likely that more rather than fewer jobs will be lost and that in 1995/6 greater demand for electricity will force buyers either to use more gas or import coal, at the expense of the locally produced coal that would otherwise be available.

Bibliography

Anderson, D: *Coal: A Pictorial History of the British Coal Industry*. David & Charles, 1982.

Ashley, M: *England in the Seventeenth Century*. Hutchinson, 1978.

Ashton, T S: *The Industrial Revolution 1760–1830*. Oxford University Press, 1968.

Ashton, T S and Sykes, J: *The Coal Industry of the Eighteenth Century*, 2nd Edition. Manchester University Press, 1964.

Ashworth, W: *A Short History of the International Economy since 1850*, 3rd Edition. Longman, 1974.

British Coal/National Coal Board: annual reports.

Caminus Energy: *Markets For Coal*. HMSO, 1993.

Chester, Sir N: *The Nationalisation of British Industry 1945–1951*. HMSO, 1975.

Childs, D: *Britain since 1945: A Political History*. Methuen, 1986.

Court, W H B: *A Concise Economic History of Britain*. Cambridge University Press, 1965.

Dalyell, T: *Misrule*. Hamish Hamilton, 1987.

DTI: *White Paper – The Prospects For Coal*. HMSO, 1993.

Ernst & Young: *Review of Organisational Efficiency and Overhead Cost Structure of the Mining Activities of British Coal Corporation*. HMSO, 1993.

Ezra, Sir D (Editor): *The Energy Debate*. Benn Technical Books, 1983.

Gray, W: *Chorographia: Or a Survey of Newcastle upon Tyne in 1649*. S Hodgson, 1813.

Gormley, J: *Battered Cherub*. Hamish Hamilton, 1982.

Gregg, P: *A Social and Economic History of Britain 1760–1972*, 7th Edition. Harrap, 1973.

Griffin, A R: *The British Coalmining Industry: Retrospect and Prospect*. Moorland Publishing, 1977.

Hall, A: *King Coal*. Penguin, 1981.

Heinemann, M: *Britain's Coal*. Victor Gollancz, 1944.

Hill, C: *Reformation to Industrial Revolution*. Pelican, 1969.

HMSO: First Report of the Commissioners on the Employment of Children in Mines. 1842.

HMSO: First Report from the Energy Committee: The Coal Industry. 1987.

HMSO: Retail Prices 1914–1990.

Hobsbawm, E J: *Industry and Empire*. Pelican, 1969.

Howe, C: *China's Economy: A Basic Guide*. Granada, 1978.

James, P: *The Future of Coal*. Macmillan, 1982.

John T Boyd Company: *Independent Review 10 Collieries Under Consultation*. HMSO, 1993.

John T Boyd Company: *Independent Analysis 21 Closure Review Collieries*. HMSO, 1993.

Kernot, C P H F: *Mining Equities: Evaluation and Trading*. Woodhead Publishing, 1991.

Lawson, W: *The View From No. 11*. Bantam Press, 1992.

MacGregor, Sir I: *The Enemies Within*. Collins, 1986.

Mathias, P: *The First Industrial Nation: An Economic History of Britain 1700–1914*. Methuen, 1969.

Mining Annual Review and *Mining Magazine*. Mining Journal Publications, 1989–1992.

Mitchell, B R and **Deane, P**: *Abstract of British Historical Statistics*. Cambridge University Press, 1962.

Nef, J U: *The Rise of the British Coal Industry*. George Routledge & Sons, 1932.

OECD: *Coal 1992*. 1992.

Pearson, P (Editor): *Prospects for British Coal*. Macmillan, 1991.

Phillips, J: *A Treatise on Inland Navigation*. 5th edition, 1805, reprinted David and Charles, 1970.

Roth, A: *Sir Harold Wilson, Yorkshire Walter Mitty*. Macdonald and Jane's, 1977.

Smith, A: *An Enquiry into the Nature and Causes of the Wealth of Nations*. Penguin, 1986.

Trade and Industry Select Committee: *British Energy Policy And The Market For Coal*. HMSO, 1993.

Turner, G: *Business in Britain*. Eyre and Spottiswoode, 1969.

Ward, C R (Editor): *Coal Geology and Coal Technology*. Blackwell Scientific Publications, 1984.

Index